THE
HEART-MIND
MATRIX

"Of the hundreds of books about evolution I have read, some of which I've written myself, this book best captures the sense of what evolution is. It takes you into a realm of celestial playfulness to provide a sense of its living presence, of what it feels like. It is like the difference between looking at evolution through a microscope, or laid out in charts on a wall, and feeling yourself within it, part of it—and able to do something about it."

DAVID LOYE,
AUTHOR OF *DARWIN'S LOST THEORY*
AND *DARWIN'S SECOND REVOLUTION*

"In 20 years of personal experience I have found Joe Pearce to be original, provocative, challenging, insightful, courageous, painfully honest, humble, and generous beyond reason—which places him in my book right alongside David Bohm, Ashley Montagu, Krishnamurti, Socrates, and other world teachers."

MICHAEL MENDIZZA,
COAUTHOR OF *MAGICAL PARENT, MAGICAL CHILD*

"The startling arc of Joe Pearce's vision has proved true time and again. We ignore him at our own peril."

LEE NICHOL, EDITOR OF *THE ESSENTIAL DAVID BOHM*

THE
HEART-MIND
MATRIX

*How the Heart Can Teach
the Mind New Ways to Think*

JOSEPH CHILTON PEARCE

Park Street Press
Rochester, Vermont • Toronto, Canada

Park Street Press
One Park Street
Rochester, Vermont 05767
www.ParkStPress.com

Text stock is SFI certified

Park Street Press is a division of Inner Traditions International

Originally published in 2010 by Goldenstone Press under the title *Strange Loops and Gestures of Creations*

Note to the reader: *This book is intended as an informational guide. The spiritual fasting techniques described herein should not be seen as an endorsement of this practice and are not recommended for the uninitiated. They should be pursued only with extreme caution.*

Library of Congress Cataloging-in-Publication Data
Pearce, Joseph Chilton.
  The heart-mind matrix : how the heart can teach the mind new ways to think / Joseph Chilton Pearce ; foreword by Robert Sardello.
    p. cm.
  Includes bibliographical references and index.
  ISBN 978-1-59477-488-1 (pbk.) — ISBN 978-1-59477-510-9 (e-book)
  1. Heart beat and intelligence. 2. Intellect. 3. Human evolution—Religious aspects.
I. Title.
  BF431.P343 2012
  158—dc23

                                                        2012007950

Printed and bound in the United States by Lake Book Manufacuring, Inc.
The text stock is SFI certified. The Sustainable Forestry Initiative® program promotes sustainable forest management.

10 9 8 7 6 5 4 3 2 1

Text design by Jack Nichols
Text layout by Virginia Scott Bowman
This book was typeset in Garamond Premier Pro with Trajan Pro and Gill Sans used as display typefaces

Figures 5.1 and 5.2 (pp. 68, 69) courtesy of SCI Institute, University of Utah
Figure 5.8 (p. 73) courtesy of Lockheed-Martin Solar and Astrophysics Laboratory (Drs. Markus Aschwanden, Marc DeRosa, and Carolus Schrijver)
Antonio Machado poem (p. 165) from *Border of a Dream: Selected Poems of Antonio Machado,* translated by Willis Barnstone, courtesy of Copper Canyon Press

Dedicated to Michael Mendizza

CONTENTS

FOREWORD

—◆—

By Robert Sardello

This book is a many-leveled, polysymphonic writing on the human spirit that can be heard on one level or many intertwining levels simultaneously. I hope to alert you to a reading style that allows the sentences to resonate within your bodily being, wherein the book has the power to heal our disturbed brains. When Joe Pearce writes that the child's most important inner directive is to maintain contact with his or her caretaker or nurturing one, and explore the world out there, he speaks to deep and forgotten recesses of the soul. We can hear a resonance going beyond those who nurtured us initially, and open again to a full soul contact with an all-nurturing Source.

The first half of this book shows how imbued we all are with the pathologies of our present day, and just noticing and becoming aware of these pathologies can be a marked advance for us. To notice, for example, how we are continually beset with subtle contradictions in social directives, such as, "Good job, hope you can keep it up." Such everyday happenings are usually passed by, but one begins to sense the contradictory opening up and cramping in of the body that ensues; or becoming aware that our attention is constantly split in dozens of different directions through habits so ingrained they seem part of the inbuilt wiring of the brain, pathologies that are given top value in our "multitasking" world; or to become aware of and notice our constant anxiety, defensiveness, and underlying fears, amid the equal offerings of ways to medicate or drug them away.

A small dimension of freedom and creativity still exists for us, seated in the prefrontal cortex of our brain, but unnoticed in the ever-present pull toward the fear- and survival-centered lower brain. Just noticing, becoming aware of our capacity for noticing itself, we find such capacity within us existing in perfect calm, a center that can re-establish coherence, an inner body harmony with the widest spiritual reaches, though typically overpowered by the negative commands of a culture of fear. Just holding to the small act of the freedom to notice, to become aware, will show, as your reading proceeds, not only what is at stake in doing so, but how this awareness can intensify and increase in resonance.

Such noticing is quite different from a self-therapeutic psychology. This noticing is of a spiritual nature based in the capacity of attention, an intention that can be very focused and yet diffused at the same time—we might say "here within" and "wide open" simultaneously.

This book addresses not just the severely critical concerns of our culture and time, but points toward our capacity and wisdom to go through the "thick of the world" and discover that right here, in the midst of chaos, is where Spirit is to be recovered. This thankfully spares us the wispy sort of spirituality that abandons the world for some supposed heights of the ecstatic. Indeed, most spirituality is not only other-world oriented, it is also anti-this world—an escapist spirituality that opens the door for the very abuses that validate our destructive culture.

So to really see what is going on in the chaotic conglomerate of culture, one has to swim in it with acute awareness and be able to recover, by deep searching and sifting for the spiritual elements present, those which provide a re-orientation from within. This project, it seems to me, is the essence of the tactic taken by this book. It comes, perhaps, from the wisdom of a life that never made the false split of spirit and world in the first place.

This may sound heavy, but no self-sacrifice or heaviness is involved in the spiritual endeavor this author has followed for decades. Rather, it comes with the insight of the spiritual nature of play—awareness that when one truly plays, one is also being played with in loving care by the spiritual Source. Free of the cultural pressures of competition with its winners and losers, here is the invitation to enter into that true play always at hand, on every level.

The whole sense of this writing originates in and expresses the intelligence of the heart, so different from a book out of the intellect in our head. *Heart-Mind Matrix* is in no way written sterilely from the spectator point of view, even as the research drawn on often is. As you read, you will feel taken into a world where everything resonates within your body, your consciousness, your soul, and spiritual awareness.

The central insight of this book—developed in many and varied ways—is the "looping" activity of fact and possibility, the given and the potential. This looping-mirroring is a living, ontological activity that, once noticed and attended, begins to awaken us from our typical hypnotic state of survival awareness and its fearful compulsions.

Although this book is essentially a spiritual writing, it does, in a certain clear sense, incorporate aspects of a materialistic science. But it turns such science inside out in a reversal of consciousness, one that can take place while reading. Sensing that almost anything seems possible to the mind's creativity, technological science assumes anything it imagines can be brought about with impunity, without regard to ethics or consideration of consequence. Becoming aware of the intelligence of heart, we are compelled to operate within a new form of science that asks not just what is possible, but what is appropriate—appropriate to the well-being of self and Earth. Such a question does not originate in the mental realm but the spiritual, and is felt bodily, once our senses and heart are attuned. So the central part of our being that simply must be allowed to function and be attended is the heart. Once you enter into the part of this writing concerning the mysteries of the physical heart, and let that text resonate bodily, engagement with the healing of brain and mind ensues. Most writing on the heart tends toward sentimental mush having little to do with the heart, but all to do with our culturally conditioned mind using emotion to render heart as a pool of froth that cannot do anything. There is, however, on the fringes, a science that works with and for the Whole, and found here are astounding findings concerning an intelligence of the whole as it emanates from the heart.

These findings can and are easily usurped by the cultural counterfeit of a true science, one which mirrors the actual while leaving the Spirit out, allowing science to be used as servant of the technology of the "do whatever is possible" variety, without regard for consequence. There is great subtlety in the ways our author negotiates these findings, bringing out the implications fully—that the heart is the center of true intelligence, and thinking that operates without this center (the binary, exclusionary, digital intellect) can neither apprehend, understand, or develop technologies of the Whole.

The discoveries of field-effects, such as the radiances of the heart, have been detected on a materialistic level by a counterfeit science. Blinded by our cultural trance, we miss the implication of nonlocal fields actually being involved with their localized expressions, as found in materialistic science. A prime example concerns the scientific findings detailed in this writing, which show the connection between the radiating field of the heart in its conjunction and resonance with the radiating resonant fields of the Earth and Sun. These findings are sound, but are based on being able to measure the electromagnetic activities of heart, Sun, and Earth. Taken for granted is that the totality of field phenomena are electromagnetic, that is, completely material. This mistaking the messenger for the message brings the blind assumption that the heart is "only physical," in spite of ages of cultures that recognized the heart as spiritual and physical simultaneously. It is an instance, found throughout present science, of taking the footprint in the snow as being its own cause rather that the result of something unseen.

We can see the electromagnetic messenger through electromagnetic instruments, but not the message conveyed, which is of the nonmaterial Spirit. Intellectual materialism, because it cannot see the source, discounts anything that cannot be measured. Taken into researches of the heart, this kind of thinking results in the formation of a "double" of the human—the self—caught in the illusion of controlling inner, invisible processes.

This book, in tracking such findings through a bumbling science, brings a restoration of such to their true Wholeness. How the author does this is an example of "remaining connected to the source of nurturing while exploring the world." The result of such a humble "method" is that the widest regions then begin to open.

Science's inability to follow this "spiritual science" may not be due to just the politics of science, but more to the conditioning we all undergo by our enculturation itself. As a culture we are caught in the throes of a massive imbalance between the dictates of our sympathetic nervous system (that automatic and natural danger-alert response of the "old brain," with its large array of attendant negative hormones and disturbances), and being unable to bathe in the restorative powers of the parasympathetic system and its balancing wholeness. Science is itself part of this culture of fear. How could it be otherwise? Scientists are not exempt from our cultural imbalances and the absence of true nurturing. They, too, live in the bombardment of a counter-active electromagnetic force; they, too, are subject to the heart-numbing force of a truncated development of self, the ultrablind then leading the blind.

Admittedly, I have slanted this forward by the introduction of the slippery word "spiritual." It sounds like starting with a conclusion, but is not. I would not have introduced the word if it did not appear as a phenomenon within the interstices everywhere within this book. It is, for example, not falling into materialism when the author points out that the "self" comes into physical embodiment in the "neural tube" of earliest embryonic life, which is the matrix of the developing brain, while at the same time is the source of the morphologically unfolding heart. If one is caught by a science without a Source, the true importance of such a finding slips by us and remains within the dusty annals of other scientific findings (unless someone comes along with a technology to use such findings to further the scientific illusion of control over all natural process). This brings our life into a double of itself, a split system. But by paying attention to these findings with all one's wisdom rather than intellect alone, one will hear what these findings are saying: there is an indivisible connection between heart and mind.

Hearing in this way, we can follow such findings out into the farthest reaches—not by "jumping" out there as though without a clue, but by full inner listening. We do not have to jump into hypothetical regions, but rather witness and begin to resonate with a capacity to see/feel/know the presence of the spiritual realms as integrally in resonance with the physical realms. This allows the author to say, with conviction and precision, that the self expresses the universal through the heart, and the individual aspect of things through the mind. Wow! The strange loop—mirroring phenomenon—is taken to the outermost reaches. We have a way, built into our very nature, of being "here" (in this flesh) and "there" (in the spirit) at the same time—and this is the way toward restoration of the true sense of culture as an expression of creation.

Joseph Pearce keeps listening, ever deeper. If we are to find the way through the paralyzing activity of living without a nourishing presence, it is finally death that must be confronted. Not theologically or spiritually in the old sense of spiritual—that is, detached from the rich thickness and substance of the tangible world that resonates with the presence of spiritual activity. For our survival mentality and its false culture is merely trying to protect itself from the threat of death.

This work addresses just this deepest of all human questions—the unknown of death—in a most creative way by addressing the phenomenon of evolution. This phenomenon also starts the book and is an inner presence throughout the central chapters. In the final chapters the issue returns in its full significance. Bye-bye false opposition between "creationists" and "evolutionists." Here in this book you will find a truly creative understanding of evolution, as an urge within all creation to overcome constraint and limitation as they arise. In humans, this includes death, and our crippling fear of it. The clues to this new vision of evolution are to be found in the mysteries of DNA as the bridging of matter and spirit, body and soul. Once moving from a material to a holy science, we see how connection with the intangible can set up a powerful resonance that can bring the fulfillment of our evolution. This kind of resonance, which can move matter from the invisible to the manifest, Pearce terms "crossovers." He tells numerous stories of how such crossovers between the physical and nonphysical have occurred with particular individuals, which leads to what might be felt as our having worked the issue to a neat and tidy point. But not so!

The real bombshell of the book is found in the manner in which the reality of Spirit is directly addressed. Though related to the phenomena of "crossovers" (as between potential and actualization), the living reality of Spirit is in a realm to itself. The living presence of Spirit in an individual, and certainly within cultures (which is in no way connected with religion or being religious), is not an issue of "wisdom-thinking" as worked through before. Resonance cannot account for the incidents of people being "infused" by Spirit. The author gives profound instances of such "Spirit-infusion," which, though moving beyond the mirroring of the strange loop category, once having occurred enter into that mirroring effect itself by intensifying the resonance of the creative within it.

So I have slightly modified my opening paragraph of this foreword by saying that much of what is talked about in the book, as the mirroring of the physical by the nonphysical, still falls short of full Spirit as itself. For Spirit is not a field at all, and cannot be encompassed by the "laws" of fields and field effects. And everything that is said here concerning the strange loop—mirroring effect, as well as resonance, leads right to this door of Spirit—and I doubt that there is any other pathway that leads so clearly through the world to this door.

The question arising at this doorway is: How can we be completely, bodily open to Spirit, that we might ourselves be infused by it, yet still tend to the everyday world? No mental answers are spelled out here, but the implied koan is given throughout these pages.

Robert Sardello, Ph.D., is the cofounder and codirector of the School of Spiritual Psychology and coeditor of Goldenstone Press. He is the author of *The Power of Soul*, *Facing the World with Soul*, *Freeing the Soul from Fear*, *Love and the Soul*, and *Love and the World*.

ACKNOWLEDGMENTS

Lee Nichol should be listed as coauthor of this little volume of mine, for certainly he was a major force in shaping its structure, making it coherent and logical, and bringing it to publishable form. (He quietly turned down my suggestion of coauthorship.) In well over a half-century of writing and working with editors, however, none has brought the caliber of logical clarity, intelligence, and capacity for literary organization, as Lee. Just to find someone who could intuit my intent and meaning in the often-jumbled hodgepodge of my writing was itself rewarding. So, in this matter of having been blest by a brilliant and splendid editor I must leave much unsaid other than, "Thank you, Lee."

This book's dedication to Michael Mendizza is a logical parallel-companion to the above praise for Lee Nichol. Michael has been a loyal and constant supporter for some twenty or more years, aiding my efforts in many different, quiet, but critically important ways. Fittingly enough, in their younger years, both Michael and Lee were involved in filming, editing, and helping in the chronicling and publications of the thoughts, writings, and general works of the physicist-philosopher David Bohm (long a hero of mine, cited in my first book, *The Crack in the Cosmic Egg,* five decades or so ago, and a serious influence since).

In addition, I am equally thankful to Robert Sardello, to whom I dedicated my last book, *The Death of Religion and Rebirth of Spirit.* Sardello is easily one of the most important figures and influences in my life, particularly in these closing years of mine. Now I stand doubly grateful to him for his generous foreword to this current effort. Rather unintentionally, perhaps, Sardello's foreword seems virtually a guide to reading the book itself, as well as a key insight to its intent and contents.

So, with these several lines of force converging here, to my good fortune and for which I am grateful, I conclude, last but not least, with acknowledgment of my good wife Karen's patient support, and the continual editorial help of my daughter Marian, a skilled and thoughtful writer in her own right. Marian's ongoing readings and critical advice have continually helped clarify where I too often and so easily become murky and unclear.

A MIRROR OF THE UNIVERSE

Evolution as a principal part of creation is a force or impulse to overcome any limitation or constraint that might arise in created beings—such as us. Further, there is no evolution except through creation, and no creation except *through* evolution, and no life at all without both.

Back in the early half of the twentieth century, scientist Arthur Eddington pondered on how it was that this three- or four-pound clump of jelly in our skull could have discovered, come to understand, manipulate, and even control so many secrets of the universe in so short a time as we have.

Eddington mused that "Man's Mind must be a mirror of the universe." Engrossed in writing my first book, *The Crack in the Cosmic Egg*, I was delighted with Eddington's imagery, which verified a principal pillar of *Crack* (begun in the late 1950s). "Why of course!" I mused in turn, as I worked out Eddington's issue. "Man's mind is a mirror of a universe *that mirrors Man's mind.*" Each, I claimed, brings about and sustains the other.

Place two mirrors directly opposite each other, I suggested, and observe the "infinite regress" resulting, as the endless series of reflections unfolds, stretching off toward an infinite nowhere point. To ask which mirror reflects first, giving rise to such replication regress is as fruitless as the issue of mind and its reality. For there is no beginning or

1

ending of such processes or the minds musing on them. Their reflecting beginning is in their reflecting ending and vice versa.

In spelling this out, I claimed that a scientist, in his passionate research leading to a great discovery out-there, has, unbeknownst to himself or his scientific community, *entered into* that discovery, as an indeterminable but integral part of the creation itself. While we do not "create" our reality or world, there is likewise no ready-made world-out-there awaiting our discoveries or creations within it. Mind and its world-creation give rise to each other, just as do creator and that-created.

My second argument has been that creation is not just an intelligent process; it is intelligence itself, as is its vital counterpart, evolution. Just as this creative-evolution doesn't "have" intelligence, but is intelligence, whatever is created therein is equally an expression of that intelligence (as will be picked up again in chapter 7).

Creation is an endless process stochastically exploring every possibility of being. *Stochasm* is from a Greek word meaning randomness with purpose, and to deny a random factor in creation-and-that-created is an error, as is denying purposefulness behind that randomness or its creation. Evolution is an ever-present pressure or urge within any and all created phenomena or living events to move beyond any limitation or constraint within such event-phenomena.

So evolution is the *transcendent* aspect of creation, rising to go beyond itself, being the response of life to its own ever unfolding evolution. And every living phenomenon or event reaches, at some point, its eventual limitation and constraint. There could, in fact, be no creation that is final, since even the concept of finality would indicate limitation against which evolution, as is its nature, would, perforce, move to creatively rise and go beyond.

To move beyond limitation and constraint is a twofold process: first, to generate such movement itself, and second, to create that which lies beyond and manifests through that movement. And that which lies beyond the limiting constraints of something created, comes about only by the movement of transcendence "going-beyond" itself.

"Where" transcendence might go in "moving beyond" is determined by the going itself, as we will further explore in chapter 11. ("We walk by falling forward, and go where we have to go . . ." as poet Roethke expresses it.) Our "going" enters into the nature of that which we enter into and brings about by our going—which is the very definition of the *strange loops* rising within this "mirroring" process. And herein lies the central thesis of this little argument of mine, as it did in my first book, *Crack in the Cosmic Egg,* well over a half-century ago.

By its nature, evolution reveals all "points of constraint-limitation" in creation, and creation takes place stochastically in response. Stochasm's purposefulness acts to select, out of that profusion, what is needed or what works. Like water, evolution seeks out, through means, which come about *through this seeking-out,* a level that lies beyond its present state. Since evolution's "present state" is always the moment of its stochastic movement to selectively go beyond itself that "moment" is ever present and all there is. As Robert Sardello points out, in that moment, the future will flow into our present, "making all things new," if allowed.

Enter Death:
The Ever-Present Third Factor
in the Mirroring of Evolution-Creation

Not death itself, but the fear it engenders, is the greatest of constraints. Moving to overcome this limitation is complex, since abolishing death (were it possible) would change the very ontological constructs of reality as it has evolved. So even the *notion* of abolishing death, as proposed in some Eastern philosophies, is counterproductive and would stop creation in its tracks. Fear involves the complexities of our ever-changing emotions. Emotion involves our capacity to relate and interact, which is a cultural issue, not ontological, and tackled later in this book. (Ontology is a general theory-explanation of how reality arises and functions, and such an idea or system of thinking proves to be as critical an issue as creation itself, not just a mind-wandering "ideology.")

In the meantime, consider that cosmos and person (you and me) are of the same order, the same essential creative function, regardless of the size or measurements involved (light-years or micrometers). We were that creative function on its "micrometer scale" until we invented the electron microscope, which calls for even finer gradations or "scalar measurements," cumbersome refinements we can do without here.

For clarity at this point, however, a certain "biological" complexity is involved. Remember that *bio* simply means life, while the *logical* way it works is our knowledge of this life process. And this process has a very logical "way it works," though not some machine-like or precise way. Wandering or meandering stochastically through its endless branching ways is more true-to-life as it happens.

So, with such precautions I launch into the following brief simplification of what is equally an explanation of this incredible intelligence within our life's design. The sample given here involves a wondrous, actual and vital-to-our-life creature called *mitochondria*.

The Unfathomable Intelligence of Creation-Evolution Found in Mirochondria, an Early Arrival in Our Matrix-Sea

Neurons, or brain-cells, are created early on in the neural tube of the fetus in the womb. The neural tube is an organ sprouting somewhat early in a fetus's body. It morphs into the heart during gestation and produces neurons en masse as it does so. These neurons are not self-mobile (as some cells are), and are actually towed from the fetus's neural tube into their eventual head-position by mitochondria (see the work of Rakic and of Sagan and Margulis).[1] Mitochondria are some of the tiniest and earliest of all life-forms, as well as an enigma that well illustrate the brilliant design within or behind creation-evolution, an enigma that needs a brief section of its own here.

At some early point of Earth's evolution, when the materials and conditions were just right, Nature birthed a vast profusion of single-cell life-forms in her matrix-sea. (From this word *matrix* we get matter, material, mother, and other sources originating in the sea.) These single cells were both nucleated and nonnucleated—with and without a DNA molecule—and among this vast profusion, those nucleated cells offered possibilities for mutation and development. Those, which best fit into or could adapt to the ever-changing environments' ebbing and flowing and could survive and/or thrive in them, evolved to ever-larger forms time and again, some finally crawling out onto land, as detailed by Darwin, and exemplifying stochasm—purposeful randomness—as Nature's way from the beginning.

Among this limitless plethora of random cells with nuclei, one particular and very tiny single-celled creature stands out. Labeled *mitochondria,* these rather box-like cells were and are exceptional and rare in having only half of the conventional DNA molecule needed for reproduction of themselves and possible mutation into a more complex form.

According to microbiologists Margulis and Sagan, these mitochondria seem to have remained pretty much the same as first created in those primal seas. Short-changed as mitochondria, with their half-DNA, they were apparently designed *not* to evolve, since they met the needs of Nature's long-range plans for the myriad life processes swimming around at that time, as well as time yet to come. On her first try, Nature apparently hit her target for what would be needed in the coming ages of evolution. This stable, unchanging, nonmutant mitochondria seems a case of a selection without the profusion from which selectivity ordinarily takes place, rather a backward way of working, as in the mirroring of a strange loop phenomenon. (Of course Nature, who has a corner on the creative market, may have run through a trillion tries before hitting on mitochondria as what she was after, since time was no factor at that point and doesn't apply. And those preliminary attempts may have simply vanished without trace, "like foam trails on the sea," as poet Machado expresses it.)

With their short-changed DNA, mitochondria can survive and

multiply only by being incorporated into other fully endowed cellular forms—and thus their very small size: they can fit into any variety of cellular creatures. On being incorporated into other cellular forms, mitochondria must first borrow a portion of their host's DNA, and this borrowed bit serves as the other (missing) half of the mitochondria's own DNA.

Through this incorporation, or adoption, mitochondria, with their now-whole DNA (half native to them and half adopted from their matrix-environment), are "read" by that host body as being a benign part of that host's body itself. Otherwise mitochondria would be "read" as an invader to be expelled by the host's immune system. Thus, mitochondria can fit into and be accepted by any creature, as that creature evolves. On being incorporated into a cellular structure, this now-completed mitochondria serves a whole laundry-list of functions critical to the life-form, an example of perfect symbiosis in an incredibly brilliant design.

Each—adoptive cellular structure and adopted mitochondria—gives rise to the other. Neither system can stand alone, as is true of humankind and Earth herself, who also seem to give rise to each other. They (and we) are independent, while yet inter-dependent.

Among the myriad critical functions served, mitochondria are the means by which protein conversion into nutrients, or "energy," takes place, to keep that cell (and itself) alive. Mitochondria reproduce or replicate themselves on demand, according to the needs of their host, and the host in turn meets the needs of her adopted and versatile mitochondria.

Neuron Migration

Most intriguing, graphic, and illustrative in this intelligent design is the critical role mitochondria plays in the migration of neurons from the neural tube into their locus in the upcoming brain of a fetus. The neural tube, one of the earliest embryonic organs, gives rise to both the brain and heart (as will be explained in chapter 7). This migration into the upcoming brain's locus in our head is brought about by mitochondria literally *towing* those new neurons into location in what will be our cranium, where our emerging brain will be housed. (Pasko Rakic's pictures of this micro-tugboat affair are intriguing.) As the neural tube morphs into the heart over the ensuing weeks, untold millions of neurons flow from that tube, and mitochondria reproduce themselves accordingly, as needed to drag, or tow, that constant flow of hatching neurons into their new matrix, the brain-to-be.

Mitochondria do this by means of slender threads, which they create on-site, for the purpose of this towing job. When the neuron's destination is reached, the connecting thread disappears and the neuron takes its place among the growing millions, according to its own destined part of the brain's structure.

What about this thread made for the purpose? From whence did it come and where does it go when it disappears? At best these are questions arising from a "misplaced concreteness," to use A. N. Whitehead's term. The threads are, like mitochondria themselves, simply a phenomenon needed at that time for a single purpose. When such a thread is no longer needed, it simply isn't there, nor, apparently are the massive numbers of mitochondria created on-site for the job. Effect can precede cause as needed in this magical mystery tour of creation, though such an observation is somewhat offensive to strict materialists.

The Tale of the Missing Tail

Mitochondria play an equally critical role in human reproduction (to which we will also return later). A sperm's long tail, by which it propels itself upstream to its destination, dutifully disappears when it reaches its goal and is no longer needed. That tail-of-the-sperm consists of nine microtubules forming a most magical circle of such tubules, and making a quite serviceable and efficient tail. Mitochondria power this magical-circular tail until—at sperm-journey's end at the portal of that life-source egg—that vigorous tail disappears.

The sperm's tail doesn't "drop off," as one might speak of the event, since there is nothing to drop. That tail has been only an oscillation of frequencies furnished by the mitochondria; the smart critter simply switches off the power when the tail is no longer needed, at which point the tail isn't.

Understanding that there is no sperm-tail that drops off, only a "vibrating" or oscillating frequency that stops its oscillation when no longer needed is a key to how life works. One is tempted to say "*it* stops *its* oscillations," but there is no "it" that oscillates—there is only oscillation as an event-function. By oscillation is meant a "turning off and on" of whatever power, radiation, force, frequency, or resonance that is involved in and brought about by mitochondria.

There is no end to oscillations oscillating, since in one sense there is no beginning either. What is "it" that would begin or stop oscillation? Oscillation, like creation, is a verb, not a noun, nor is there much to reify here and give substance to, even if "only" imaginary.

Tubulin and the Stuff of Microtubules

We speak of nine microtubules in a circle forming a tail, but what is a microtubule? Every cell of our body encases that folded-up bundle of DNA and a cellular "filling" of sorts called tubulin. Tubulin, in turn, is a kind of semantic substitute for an unknown and probably unknowable function-process—giving it, if not some "reified" substance, at least a label to which we can refer. (Reification means to attribute substance, or material stuff or "thingness," to some imaginary wish-think product of our mind. This, as physicist David Bohm insisted, is an error of our thinking that can and too often does lead us astray.) Tubulin is that of which microtubules are made, as microtubules are made of tubulin, surely a trick of a mirroring tautology. To say that tubulin is not a substance as ordinarily spoken of, but rather an oscillation of that same mitochondrial-sourced "power" that forms the sperm-tail and powers it on its journey, is apparently much ado about nothing. Yet it is a key to everything.

Such oscillation may be, in fact, similar to the "quanta" of quantum theory, wherein we find a quantum has no final substance, but "ends" as such only in relationships or resonance among other things (at which point we exercise a bit of reification to give such abstract nothingness a bit of "substance," at least for discussion and perhaps peace of mind). Tubulin and its microtubules, without which cells could not exist, give a causal basis for an event that is otherwise inexplicable to our logical, rational "scientific" mind; they are "effects" without a cause, rather a no-no in earlier science.

We should as well consider Mae-Wan Ho's observation that a cell is "made" of a liquid crystalline form-substance, similar to the screens of televisions and computers. From one standpoint, Mae-Wan claims, the body's seventy-eight trillion cells function as a single, coherent liquid crystal. This observation changes the complexion of nearly everything in physiology-biology while highlighting the role of the many mystery-characters behind the scenes in our life, functions that work perfectly well without our knowing or needing to know much about them.[2]

So in summary, mitochondria are known to be the power-providers of cellular life, transforming proteins into usable energy and vice versa, as well as aiding in a host of other critical tasks, such as the discovery by Rakic that mitochondria aid in bringing about neural cell migration from the neural tube. And those neurons pour out of the neural tube by the millions, as that heart morphs into old-thumper; and to this morphing, mitochondria respond by multiplying themselves accordingly, spinning out more mitochondria and those threads with which to drag those neurons into their new home in our head. And herein, to my thinking, lies a significant indication of an incredibly intelligent design functioning alongside, as an integral part of, the stochastic random-chance-selectivity of evolution, where they seem virtually a creation *ex nihilo*. That is, *out of nothing at all*—this *ex nihilo* being a no-no in academic science, *but what a wondrously enchanting nothing-at-all it proves to be.*

ONE

---·---

THE FALL OF MAN

A humankind abandoned in its earliest formative stage
becomes its own greatest threat to its survival.

MARIA MONTESSORI, M.D.

Holding in mind the well-known and well-worn issue of Charles Darwin's first opus, *The Origin of Species,* based on genetic mutation, selectivity, and survival of the fittest of those mutations, we will here consider his equally great second, ignored, and almost unknown work, *The Descent of Man.* In this second work, representing the later and more mature half of Darwin's life, he shows how humankind arose through the "higher agencies" of love and altruism. Selectivity and survival, being foundational, are retained, but in service of this higher and more complex life-form.

The issue of this higher evolutionary cycle found in *The Descent of Man* (hereafter referred to as "Darwin II") lies with nurturing, which instinct gives rise to, fosters, and allows love and altruism. This aspect of Darwin's life-work, as clearly articulated by David Loye's excellent little volume, *Darwin's Lost Theory of Love,* has been grievously ignored. Consider some of Darwin's essays and interests of this later period, such as "Selectivity in Relation to Sex and Expressions of the Emotions in Man and Animal," and his interest in the work of Ernest Haeckel, Lamarck, and Goethe. And how does one account for such ignored sentiments found in Darwin II as, ". . . aiding the weak to survive . . . the instinct of sympathy, the noblest part of our nature . . .."? Where now a species of tooth and claw, and why have such negative myths been accepted while ignoring the deeper, more spiritually inclined aspects of Darwin's genius?

Nurturing proves to be not only the way by which this human species arose out of its animal ancestry; it proves to be the only way by which we evolved creatures can then be fully developed, from conception to maturity. Nurturing is the staff and stuff of human life, the one indispensable necessity, yet now having become so rare.

Our most ancient origins, as found in *The Origin of Species* (hereafter, "Darwin I"), can be graphically traced in the parade of skulls preceding us (see figure 1.1). We were announced quite visibly, so to speak, some brief forty to fifty thousand years ago, by the appearance of a brain case with an abnormally large area directly behind the orbit of the eyes, in the frontal-most part of our skull. (Observe the striking difference of profiles of the earlier Neanderthal and later Cro-Magnon, assumed to be the first fully human creature.) This bulging forehead houses a prefrontal cortex setting us off quite markedly from all our forebears. This particular neural grouping constitutes the high point of evolution to date, a large addition to the threefold structure found in the heads of our nearest evolutionary kin, such as the Neanderthals.

Darwin II attributes this addition to the nurturing of love and altruism.

Nurturing both gave rise to, and is the combined effect of, love and altruism—which in this book we will come to understand as a typical "mirroring" or "strange loop" effect. Whatever its roots, nurturing sustained us for millennia and what should have been human life thereafter, with the sky the limit to what such a system could do.

Paradise Lost

Somewhere along the way, however, nurturing was compromised. It was diluted and adulterated to the point of being sidetracked, insignificant to the point that it finally lost out to survival concerns, to varying extent and in different climes and times. Today, nurturing, as needed by our species and the Earth, has all but disappeared. Pockets of nurturing remained, even into the mid-twentieth century, in a few remote and isolated human groups.[1] The societies they depict offer us critically needed models to study and examples to emulate, if we are to recoup our loss of this major evolutionary tool, as nurturing proves to be.

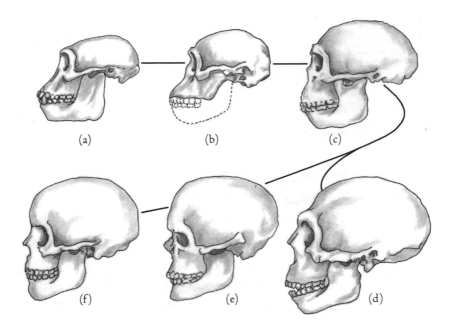

(a) Ardipithecus ramidus (b) Australopithicus africanus
(c) Homo erectus (d) Neanderthal (e) Cro-Magnon
(f) Homo sapiens sapiens

Figure 1.1. Updated variation of Ashley Montagu's profiles of skulls. The profiles show the shift of a hind-brain to prefrontal cortex, through major evolutionary periods, culminating in Cro-Magnon man 40,000 to 50,000 years ago. Note the difference between Neanderthal and Cro-Magnon, whose appearance was short-lived but gave rise to a far more permanent and stable form in modern humans. The most recent expanded skull houses the prefrontal cortex—evolution's highest point of neural development and her jumping-off point for the mind's expansion into worlds beyond. (Line drawing by Eva Casey)

Mirror to Mirror

Which Reflection Comes First?

The reason such an evolutionary setback as loss of nurturing became near-permanent wherever it occurred, is simple, ironic, and can happen rapidly. A prime example is Jane Goodall's account of a rogue ape, victim of a failure of nurturing, who upset the social structure of his whole ape-troop.[2] In the case of our unnurtured humankind, and in spite of our vastly superior brain, we were caught up in a deluge of self-inflicted disasters of every description, multiplying at every level, that followed on the heels of our nurturing failure. The irony is that the intensity of our crisis and its near-permanent status thereafter arises from, and can be attributed to, our very "superior brain." It takes some extraordinary brilliance and creativity to make the incredible mess we have made on this good Earth.

Precisely as Maria Montessori warns, we were so immediately absorbed in surviving the results of our own reactive patterns—brought on by failure of nurturing—that we had no time, energy, or interest to reflect on how or what happened, or was happening, to us. This is our condition today, where such loss and projection onto others "out there" of the causes of such loss, have been replicated age by age. Our survival concerns have greatly expanded and changed with the times, since the sharper this new intellect of ours, the deeper our crisis.

And we are getting correspondingly smarter intellectually while less intelligent. *Intellect,* a head-based operation incorporating ever more complex variations and applications, each needing further explications and qualifications, has become separated from *intelligence*—the automatic and natural state of the heart that brings coherence.

The Cultural Counterfeits

A counterfeit is a duplication of an original, from crude resemblance to those so nearly exact as to defy all but the trained observer. But no matter how apparently perfect the counterfeit is, minor, near-insignificant differences are always present, and will eventually bring ever greater problems in application. Meanwhile, the most minor miss in the fit multiplies into a major one through continual use.

To grind on this a bit, failure to nurture expresses in such a myriad of constantly branching critical problems that all objectivity suggesting a possible cause is lost in the mounting dysfunction. This leaves us aware only of the dysfunction, which by then is considered natural, or "the human condition." I spelled this out in the chapter, "Time Bomb in the Delivery Room," in my 1977 book, *Magical Child*. This effort did nothing to counter the effects of that delivery room, or time bomb, years down the road, culture being the power it is.

The importance and significance of nurturing as a survival response has, on its loss, brought in its place a mass of cultural counterfeits of nurturing. These counterfeits are "head-based" intellectual conclusions bringing roughly approximate solutions for the missing intelligence; consequently, the inevitable problems inherent in counterfeits eventually appear and absorb our attention. Origins are forgotten. And, though these counterfeits continually betray us, we are constantly seduced by them because of our fundamental needs for nurturing, with which these counterfeits have some vague resonance. Virtual reality, in its myriad of current expressions such as television, computers, and electronic stimuli of endless variety, has almost completely replaced reality as the state of live, direct biological awareness and experience as developed over millennia.

Caught up in trying to make these counterfeits work, such attempts sustain and increase the counterfeit incentives and their power. And those counterfeits, products of our ever-new and ever-sharper intellect, can border on genius itself, although always causing problems at some point, spinning our webs of error and production of counterfeits ever tighter.

Even if these counterfeit structures are analyzed and brought to light, such analyses can only be interpreted through our cultural mind-set. This mind-set automatically counters any possible conflict to itself by its own cultural counterfeit of the analysis itself. This is a largely nonconscious response on our part, simply our mind-set interpreting the information—as it must, in order to maintain itself. We, with a sigh of relief, thankfully accept our culture's counterfeit as the obvious solution for us, thus nullifying any threat of change to the culture or to our mind-set, while locking us into ever deeper disaster. In just such ways, culture is a self-sustaining field-effect with our rationale at its service.

Continually lost in making corrections in our counterfeit world, we cannot regain that benevolence-driven mind-set, which was and is our greatest survival asset and most important feature of being human. In our compulsion to right a fundamentally flawed logical worldview, we lose our connections with and ability to open to the intelligence called for—which is heart-based, not head-based. Long repetitive usage of a flawed logical approach can change brain organization to the point we can become neurally insufficient to the task of seeing the errors in our preoccupation with our counterfeits. Losing access to intelligence, we are left with plenty of "smarts" to maintain the counterfeits, none for opening to the intelligence that could reveal them for what they are.

Our razor-sharp intellect can create and build atom bombs and destroy the very atmosphere of our Earth, but the basic intelligence needed to grasp this fundamental problem of loss of nurturing *is gained only by brain-heart development itself.* And brain-heart development is a major thrust of the nurturing function itself, which is, in turn, dependent on brain-heart development. In this reciprocal and recursive movement, we find a "strange loop" in which nurturing and brain-heart give rise to each other.

Nurturing should have opened ever-new evolutionary pathways—and still could. Instead, we have locked into a survival mode, which is now considered to be not just the norm, but the "human condition" and/or "human nature." Around and through our automatic survival response we invent an incredibly complex and nonviable environment we must then attend with our whole being, both to survive in such an environment individually, and to maintain that very counterfeit environment itself, whose loss is sensed as a major threat to that worldview we share as our cultural basis, trapping ourselves at every turn.

The failure to nurture results in serious brain-mind alterations, such that any moral-ethical persuasions concerning nurturing become useless, since not really heard. We can hear only that for which we have a receptive capacity. We have had love preached to us for at least two thousand years with virtually no appreciable decrease in violence nor increase in love. Only the state of love can hear that with which it is resonant. This is a classic double bind, a Catch-22. An alternate approach—one I have long promoted—is a straight biological-neurological one that arises from a Darwin II position. In such an approach, the starting point lies in grasping the fourfold nature of our "evolutionary" brain.

History of Our Internal Civil War

One of the earliest neuroscientists to realize that this complex brain structure of ours had evolved out of and from earlier creatures was Paul MacLean, more than half a century ago. MacLean paved the way for our discovery that we have within our skull not one, but four distinct and essentially separate, interactive neural systems, developed over four distinct evolutionary periods. Through the appropriate integration, possible only through nurturing, these four neural systems cooperate as an integrated function in alignment with the heart. Should this fourfold integration fail, we become not only a seriously split system, but brain's intimate connection with heart is seriously compromised—resulting in the aforementioned "human condition."

Split between these basic evolutionary drives, and largely isolated from the intelligence of the heart, we do indeed end up at war with ourselves, individually and socially, each of us with self, split between the intelligence of the heart and a fragmented brain-mind. Self, identified with mind-brain, locks into its most primary survival systems, with culture and its violence reigning supreme.

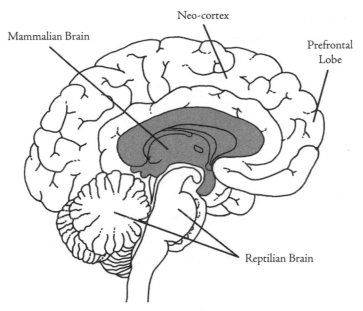

Figure 1.2. Reptilian brain without mammalian brain. Both within neo-cortex, along with prefrontal cortex. (Line drawing by Eva Casey)

The Four Evolutionary Stages

In briefest brief, the four evolutionary neural systems in our head, in their respective evolutionary order (see figures 1.2 and 1.3) are as follows: First is the so-called reptilian or hind brain, most powerful biologically since the oldest and most firmly entrenched through usage. Karl Pribram described this sensory-motor system as the "world-brain," the primordial capacity to sense our physical world and the ability to respond to that world intelligently enough to survive in it.

The next evolutionary step up the ladder was the "old mammalian" brain, built "on top of" and thus outside of our sensory-motor foundation. Through this second brain humankind could be aware of, "witness," and interpret the nature of relations between us sensing creatures and the objects or events encountered by our "world-brain" as "out there." Otherwise, we could only react to such physical events to avoid or devour them, as would the older reptilian "hind" brain on its own. This old mammalian capacity to interpret gave rise to evermore sophisticated forms of relating. For, as this book attempts to illustrate throughout, all creation is relational. Nothing stands alone. Everything is, only through relationship with something other-to-it.

So the reptilian brain gave content to which the "old" mammalian system could relate, and out of such a relationship evolved an emotional (relational) structure giving rise to relations with meaning and significance and group or herd experience—a new mammalian brain that could project beyond the sensory brain entirely. This "new mammalian" brain gave the foundation for the eventual rudiments of thought and speech through which we could reflect on an event in an abstract way; that is, extract out of direct sensory contact a mental and objective viewpoint *of* such relations.

This new capacity, in turn, made necessary and gave the foundation for a fourth, "governing" or coordinating brain to organize into a unit or single system these three independent foundational systems (reptilian sensory-motor, mammalian-emotional, and mammalian-rudimentary thinking brain.) This fourth, coordinating, and governing body of neurons (called the *prefrontal cortex*) emerged when conditions were right: sufficient maturity of the three preceding brains, and the appropriate nurturing. Once established, this new "governing" fourth brain could bring about adaptation to extremes of climate and physical conditions, and overcome the limitations and constraints that might arise. Evolution was taking wings.

Creation and the Prefrontals

Through this fourth, newest, largest, and superior brain, the prefrontal cortex, nature could organize all capacities inherent within the first three systems and bring them into a coherent, focused attention and response. Above all, this fourth brain was (is) capable of creation, in a literal sense: first creating internal images not present in the outer world (creative imaginations), then concretizing—making real—such

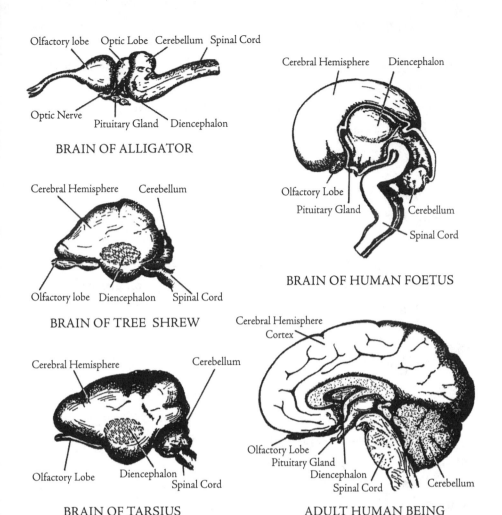

Olfactory lobe Optic Lobe Cerebellum Spinal Cord

Optic Nerve
Pituitary Gland Diencephalon

BRAIN OF ALLIGATOR

Cerebral Hemisphere Cerebellum

Olfactory lobe Diencephalon Spinal Cord

BRAIN OF TREE SHREW

Cerebral Hemisphere Diencephalon

Olfactory Lobe
Pituitary Gland Cerebellum
Spinal Cord

BRAIN OF HUMAN FOETUS

Cerebral Hemisphere Cerebellum

Olfactory Lobe Diencephalon Spinal Cord

BRAIN OF TARSIUS

Cerebral Hemisphere
Cortex

Olfactory Lobe
Pituitary Gland
Diencephalon Cerebellum
Spinal Cord

ADULT HUMAN BEING

Figure 1.3. Montagu's developmental sketch shows new neural systems added to previously developed ones, illustrating how all new evolutionary systems are built on the primary "reptilian" brain. The older reptilian brain gives rise to newer formations, which then reincorporate the older brain structures. In like manner, each successive stage evolves to overcome the limitations and constraints of the earlier structure—on which the new is always built. In this incorporating process, the newer brain transforms and expands the older structures through bringing them into service to the newer brain itself. Only through this incorporation process can the older brain rise to this function—it was not equipped to do so until the very moment of incorporation. The transformed older system thus augments the further transformation of the newer brain into its new evolutionary stature and capacity. This is a clear example of the "strange loop," showing how old and new lift one another into higher creative capacities—the never-ending story of evolution.

inner images in the "outer" sensory world shared with others (Concrete Operational Thinking, to use Jean Piaget's term).

This, in turn, prepared for, and brought about the possibility for, Formal Operations of mind, as Piaget called them, wherein we can think and imagine outside all concrete, material realms, and, in effect, operate on our very own brain structure, from which mind and such operations emerge. Nobel laureate Roger Sperry referred to this as mind emerging from its matrix—brain—and this emergent system can then bend back on, analyze, and change that very matrix giving rise to it. (A wild analogy might be of an automobile coming out of a complex assembly line and turning around to rearrange the complex of assemblies giving rise to itself.)

Transcendence, Incorporation, and Integration

So we humans evolved through a series of new neural-physical structures, each developed to go beyond the limitations and constraints of those coming before and giving rise to it. This is the essential meaning of the term *transcendence* in the present work. Yet each new system is critically dependent on the older as its foundation (a sensory-motor system is handy, after all). Though we transcend this older system we must utilize it, put it to work for us, to function at all. And in the process of utilizing any previous system, the earlier system is itself transformed by the higher system with which it must be integrated, in order for the lower to serve that higher system. This is surely both a "strange loop" and the last word in tautologies, but also allows us to say the "reptilian" or sensory-motor brain in our head is light-years beyond essentially the same component parts in the brain of blacksnake in our wood pile—while a spark plug that fires the Model T Ford is of the same general construction as that plug sparking a multi-cylinder Rolls Royce.

By this incorporation-integration, the older brain function becomes compatible with the newer, and then can serve that newer system into which it has integrated. Then the older serves the newer with those older functions, which the newer doesn't have, except *through* such incorporation of the older. Nature established this strange loop as the basis of her creative system. This system of incorporation-integration carries right on up to that frontal and prefrontal cortex, our highest evolutionary brains. The transformation the new brain brings about in its older brain maintains, enhances, and completes the older, earlier foundational capacities, as needed by the newer system to complete its own structure.

This is a procedure truly worth a bit of study because here we will find why a dysfunctional culture such as we suffer today cannot self-correct, or even be corrected, but always and inevitably replicates itself and all its errors in one form or another until finally self-destructing. In sum, the higher brain function enhances the lower with capacities, which complete the lower and enable it to work cooperatively with the higher; the higher brain must have those capacities the lower has then developed.

This incorporation-integration process plays a profound role in child development and could lead us into a rediscovery of our true nature, as co-creators in our life. Above all, here we find why abandonment of the infant-child in its earliest formative stage is such an ongoing disaster. A broken development on its own always replicates its own broken nature, in spite of all efforts to repair itself, or even attempts by others to repair it.

Strange Loops Looping

A critical series of interacting loops take place between these evolutionary brain parts—moving forward, doubling back to pick up elements or capacities needed—but *unavailable until such movement forward and doubling back* takes place. Follow this loop: the newer brain system incorporates the older into its service, giving the foundation on which the newer is built, and this incorporation transforms the nature of the older into enough resonance with the newer to function synchronously in coherence with that newer, while enhancing the original capacities of that older system. All of this reciprocal interaction is then integrated into the newer system and its supportive older system for a more advanced dual function—a new situation impacting the entire brain system developed to that point.

Our greatest great being said, "If I be lifted up I draw all mankind toward me," a statement that is an ontological truth far transcending the petty religious applications that have so misconstrued most of that great being's observations. As this "mythical He" draws all mankind toward him, he is himself transformed anew in typical symbiosis, and can thus quite truthfully state, "I am always becoming what you have need of me to be," through just this continual growth-expansion of both "sides" of the loop. By the generational love and recognition that mythical figure generates in us, that mythical figure is, himself, continually "lifted up"—giving his mythical overlaid imagery greater power to attract and lift up, round and round.

The transformation of an older primary brain system into a newly emerging one, which is, in turn, dependent on the older for its own development and transformation, is not only an example of the strange loop phenomenon, but is a major key to our evolution. For this ongoing, continual evolutionary process moves to go beyond the limitations and constraints of any current state, as such limitations arise. This transcendent action itself gives rise to new states and evolutionary possibilities not necessarily even extant before, while eventually seeking out the limitations and constraints even then becoming apparent, and kicking off another round of evolution. Neurologist Frank Wilson's brilliant analysis of the evolutionary implications of the human hand and brain shows a perfect symbiosis, each bringing more and more potential ability into and out of the other.[3] The same is true of mind and heart, as we will see in subsequent chapters.

Herein the transformation of mind itself unfolds, and we see evolution as not only an ongoing, endless process, but always a win-win situation—or at least to the extent it is allowed to be, since designed to be so if allowed. It is up to us to allow, or to impede.

This highly condensed and abstract survey ignores, for now, among many issues, the intriguing aspect of selectivity out of a randomly produced profusion. Such selectivity out of random profusion is another hallmark of our open-ended evolutionary system. And, our reviewing of this reciprocal function (as in this chapter) in a continual doubling-back on-and-of this basic material, puts this basic material into a continually wider, more inclusive basis, relating it with other aspects of evolution.

Integration and Its Failure in Development

Large and critical overlaps take place between the appearance and full-scale development of each of these four neural systems, as we unfold from conception (see figure 1.2, page 17).

So development of the first, "reptilian" brain system continues, at a more relaxed pace, long after the next, old-mammalian brain begins its development. This overlap brings a period of integration of the two systems, each supporting and influencing the other in typical, strange loop reciprocal fashion.

That is why failure or compromise of development in one stage can damage or compromise subsequent development of both predecessor and successor stages. In sum, the first neural system, dependent on integration into the second for its own completion, will be faulty if the second is faulty; and, being the foundation on which the second is critically dependent, the two "go down together," so to speak, another—and sobering—aspect of the mirroring-loop phenomenon. Failure within either system means both lose to some extent, the equivalent of a strange loop being prevented from completing its loop.

Now we can grasp the full insight in Maria Montessori's statement about abandonment in the earliest development period. An infant born and not nurtured, particularly in that most critical first year, will remain largely locked into his primary defensive sensory-motor brain, since his emotional system is then truncated, and events in his environment cannot be fully integrated. Instead, these events bring further retreats into his defensive system. He "armors" against his world from the beginning, and development will be slow and faltering. Rather than relating to the events of his world—the only way brain-mind can develop—he defends against those worldly events, increasing his isolation.

Sketching In This Fall of Man

Though many underlying factors are involved, a single major cause lies behind the mishap: our undeveloped emotional system remains embedded in and compromised by the survival system on which the emotional system depends for its own development. This survival system is still trying to establish its genetically determined functions, which it can only do through a developed emotional system, and this emotional system itself can only be completed by a functional survival system—a typical double bind. Even after the second neural system (the old mammalian or emotional) opens for its development, most of the infant's energy-attention defends against such openness.

Instead of transforming that reptilian survival instinct into a preliminary form of emotional intelligence, the higher system (the intended emotional intelligence) will be incorporated into the lower and become, in effect, a part of that lower defense system. With a compromised or damaged reptilian and old mammalian system, the third evolutionary system—the new mammalian brain—on its opening for development, will face an obvious double or triple jeopardy.

We can follow such a stage-by-stage compromise-breakdown, as it has indeed taken place historically and within us individually, until we finally get to the point where the highest, fourth brain is incorporated in whole or in part into the first, sensory-survival mode, the whole evolutionary movement thereby inverted, turned upside down, with humankind the greatest danger to humankind.

This fourth and latest brain, the prefrontal cortex, is designed to be the "governor" of the whole system—to organize the whole array into concentrations of attention-force that can carry an experience of self beyond its limitations and constraints, and move on into new evolutionary levels. But, should these ongoing and progressive compromises of nature's plan as noted here take place (which is the general case), the "human condition" becomes this compromised and crippled system, the "governor deposed," riffraff in charge, wherein our highest evolutionary potential is bent back into the service of the lowest in a de-evolutionary setback—which well describes our current massive pileup of global-social chaos.

The Breakdown of Reason

Once this reversal happens, we, still locked into our survival system, automatically employ the logical reasoning capacities of our highest brain to rationalize the violent behaviors continually erupting out of our incomplete and unintegrated survival brain. We generally defend, with all logic available to us, such destructive actions on our part as *critically necessary,* and such necessity is surely the mother of our demonic inventions.

Thus, as our history shows, there is nothing more dangerous and destructive than a brilliant, genius-level reptile, of which we seem to breed quite a number, from heads of state and industry on down the social scale to the all too similar but unsuccessful criminal behind bars. That is to say, our successful criminals are behind the corporate, political, social-scheming desks ruling our world. And therein they can justify any atrocity as deemed necessary to their success, though always cloaked as supposedly for our survival—individually or nationally. The reason we must batten on our brother's blood, imprison more and more of our fellows each year, build more bombs, go to war on and on, is "perfectly clear and logical"—over and over, blood bath after blood bath ad infinitum and nauseam. Necessity and pragmatic common sense demand it! (And, lest we forget, author and reader alike participate in this very process to varying degrees.)

Modern Man, Reptilian Hybrid

Thus, we come to our present self-destructing and nonviable devolutionary state. As inheritors of the highest brain, which evolution has produced, this lofty point is generally incorporated, unbeknownst, into the service of our lowest, sensory-motor survival system. We then function as the highest in service of the lowest, instead of *being* the highest, served by the lowest, as nature intended. Further, we then identify with these lower functions by default, and compulsively rationalize the irrational and destructive actions of these primal base reactions of ours, in an attempt to justify and maintain our mind-set, "worldview," identity, and self-esteem, thereby assuming some quasi-coherence in a most incoherent mental turmoil. The "fall of man" becomes instinctual, repetitive, circular—a demonic tautology. Or, as Walt Whitman put it, the murder comes back most often to the murderer.

TWO

——◆——

EMOTION IN EVOLUTION

In 1998 a one-page scientific article[1] stated that the emotional state of a pregnant animal entered as a determinant in the form, structure, and functioning of the brain forming in the embryo-fetus-infant in her womb. If the mother herself is given a safe, protective environment, free of anxiety and threat (or, if she can create and maintain such state within herself, as humans can), her infant will be born with an enlarged fore-brain and a reduced hind-brain. If the expectant mother feels in a harsh, unsafe, anxiety-ridden, or threatening environment, her infant will be born with a reduced fore-brain, larger hind-brain, and a larger skeletal structure and muscular mass.

Although the item received little attention at the time, the significance of it cannot be overstated, since it gives a major clue to the rising tide of violence threatening our culture and world. While supportive research has steadily appeared from several different sources over the past decade, for it to be fully comprehended and appreciated here, we need an explanation of "hind" and "fore" brains. This, in turn, calls for a brief review of infant brain growth in general, as surveyed in the previous chapter.

The "hind" or "reptilian" brain refers to the earliest neural structure developed in evolutionary history. It was carried over by Nature as the foundation of all mammalian brains, where it forms and grows in the first trimester of pregnancy in humans. The primary purpose of this hind-brain is to respond to its environment, and survive in it—both procedures being complex and lengthy. In turn, this hind-brain is the foundation of our far more advanced "fore-brain," with its old and new "mammalian" brains growing in the second and third trimesters leading to birth.

Following birth we have what is now called the "fourth trimester," bringing the growth of a fourth brain, the prefrontal cortex. This prefrontal cortex opens us, step by step, to vastly greater and more powerful vistas of human destiny. At the same time it is, as Allan Schore makes clear, both more fragile and difficult to establish than any other neural systems, and takes much longer to develop. If successfully developed, however, this prefrontal system is far more powerful than all the rest of our body-brain put together.

The Form and Content of Genetic Blueprints

Our genetic system, which guides the cellular construction of our body-brain, is just a *blueprint,* or outline sketch for such a building project; the content needed to actualize, "realize," or fill in this plan must come from the environment in which this new life is forming. Such environments or "matrices" for this new life unfold in the same sequence as the brain systems they serve: first matrix is our mother's womb, then her arms and breast, then family, society, on to our great mother Earth herself.

Nature's *form* or "outline" for this growth is given within, while its needed content is given from without. This stability of form, and open flexibility for content, leads to our eventual ability to explore almost any conceivable environment, build a "structure of knowledge" or the neural patterns of that environment, and adapt to it through such neural patterns.

A fertilized egg in a womb (holding the given DNA-RNA blueprint for new life) is thus extremely sensitive to the environment given it through both womb and then "mother and breasts," family and so on. These environments hold the content to be selected out as resonant with DNA's pattern in its ongoing growth, and determine the emotional state (or relational ability) to which the development is subject. This sensitive "outline-building guide" determines and follows the sequential stages of growth from hind- to fore-brain, sketched in above. Such growth involves a bewildering complex of cross-indexing of these form-content intricacies.

An ironic double snag appears early on, however. Our first line of defense—the reptilian or hind-brain—can be finally completed and able to function as a balanced part of a coherent, intelligent system such as ours, only as *integrated into* and *transformed by* our higher fore-brain; this is a matter of matching resonances (as pointed out previously, that reptilian system is then lifted into the lofty realm of far more advanced systems than black-snake in our woodpile). But this fore-brain must have the hind-brain as *its* firm foundation needed to establish that fore-brain itself. A potential double-bind of serious significance shapes up, while yet giving us a splendid example of the "strange loop mirroring" imperative to all growth.

Nature, in her intelligence, prefers to build her most important house of intellect-intelligence (our later-appearing neural systems) on a rock, not on sand. And while that ancient reptilian brain is a sturdy rock indeed, it must have sufficient nurturing and care to develop its rock-like character. Such care brings positive emotional hormones, which this firm foundation must have in utero, and even more as this foundational brain begins to function after birth.

Since that reptilian hind-brain is the first to grow and develop, it is the first that is ready to go to work at birth to further its development as this foundation, all depending on the emotional states involved. Those hormones arising from nurturing and care are determined, of course, by the mother, in all cases.

Being flooded with negative hormones at any part of this initial foundation registers on the new infant as tantamount to abandonment, and can compromise and seriously warp that first formation, as well as the subsequent ones following birth, which are equally dependent on this primary formation.

So this ancient and stable primary hind-brain must itself be nurtured, provided with positive emotional molecules, and established well enough to be the foundation for the second, "old mammalian brain," which itself must be developed enough to integrate this primary hind-brain into its second system—which must have the first as its foundation. Further, such integration by the second brain is necessary to the completion of the first, primary brain on which the second brain depends. This "don't go near the water until you can swim" stipulation is truly a bewildering looping back and forth, but is a clear example of the strange loop effect underlying our growth, and clearly shows why nurturing from the beginning of life is so critical.

The second, emotional-relational (old mammalian) brain begins its initial growth even as the first sensory-motor system is completing its growth. This critical overlapping period is for just such interaction-integration between reptilian and mammalian, as needed by this sequential series of interdependent building blocks. And this mutually dependent overlapping will continue throughout development, the entire fabric dependent on the appropriate emotional support (see developmental stages in figure 2.1).

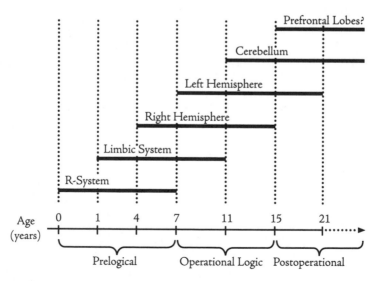

Figure 2.1. Brain development—Growth spurts and shifts in concentration

Candace Pert and Wrinkles in Our Genetic Plans

Since the mother's emotional state determines the environment the DNA must selectively prepare for, in a *negative* emotional environment adaptation selects for a stronger, larger, and more efficient survival-defensive (hind) brain and a larger body and muscle mass to go with it. Nature senses she must "armor" this new life against any inherent danger coming up in its new world.

The reverse is also true, of course. In a positive environment given by an emotionally secure mother, nature "selects" toward a stronger fore-brain and intelligence, and puts less investment in that hind-brain and its corresponding larger skeletal frame and muscular mass.

Evolution's original intent is not to equip us humans to wrestle with "Saber Tooth," but to outwit him, and she gives, or tries to give us the neural tools to do just that, which is the story of human development and the role of nurturing.

Candace Pert's early work, *The Molecules of Emotion,* clearly shows that emotion is not some wispy, nonsubstantial imagination of a neurotic mind, but a powerful hormonal element affecting, directly or indirectly, every facet of body-brain, from our immune system to the highest capacities of mind.[2]

Nature also designed our human gene to unfold, as common sense would dictate, in an appropriate sequence and *balance* of all potential systems contained in our heritage. (The nursing infant hardly needs a full set of chompers, which dutifully wait until called for.) This appropriate sequencing should give a balanced and coherent neural structure leading to lifelong development, learning, and creation. But when survival needs dominate the mother (indicated by emotional stress, anxiety, worry, and tension), balance is put on hold and a necessarily smaller percentage of that "building material" will be allotted to this later-forming fore-brain, with more going to the earlier defensive hind-brain.

Were this compensation to be carried out in all areas of development, it would neatly turn evolution upside down.

Survival is always Nature's primary concern, and concentrating her major attention on that first line of defense, all "higher" possibilities are, by comparison, "luxuries" to be afforded only after that home base of survival is secured. This is only common sense. You can't build much of a fragile neural superstructure on a weak, insubstantial foundation (again, as indicated in figure 1.3, page 19). And we need to dwell on these overlaps until some clarity regarding the entire complex is grasped. Such *looping back* and *leaping forward* can help us to establish the needed clarity.

So even before the reptilian brain is completed, the second or "old mammalian" brain begins its unfolding in the second trimester. Such unfolding establishes the foundation for further growth by a looping back of this second onto the first brain system. This happens automatically, since the second brain is built squarely on and around the first, and dependent on it structurally. Through interaction of the two, this higher intelligence will integrate enough of the older or lower into its development to simultaneously develop the older sufficiently that such integration can take place—again, a matter of resonance wherein both are "lifted up" into a new order of functioning.

Neuroscientist Paul MacLean points out that our brain functions through resonance, not objective matching or combining of content. Resonance is an underlying pervasive *quality* bringing a felt-relationship or feeling-tone of similarity, as of a shared origin, purpose, goal, or intent. Such resonance transforms the older system, which can then provide the materials for the higher or newer system to complete its unfolding. This unfolding is a matter of matching resonances. Such matching will, in turn, complete each system in its link of the sequential unfolding, and bring an ever-enlarging and strengthening of each, as foundation for the next. All of this establishes the hind-brain as the strong foundation needed for development of the whole sequence, even as the "looping" involved is repeated in all growth periods to come.

Emphasis on that fore-brain and the prefrontal cortex it gives rise to, if sequentially nurtured and developed, brings, we might say, that mythical *Peaceable Kingdom,* which artist Edward Hicks depicted in 1846, wherein lion and lamb lie down together, the hind-brain taking the "hindmost," the fore-brain the "foremost," as nature intended. But interfered with by negative hormones from any source, the reverse too often occurs, wherein that hind-brain is emphasized, dominant, and foremost (as a negative culture eventually ensures). Therein lion simply gobbles up lamb, a common survival-of-the-fittest response applauded by all (except the lamb) and awarded the winnings. And (to touch briefly on the nature of a negative culture as currently suffered) don't blame Nature, Darwin, DNA, or gremlins from flying saucers. Blame the structures of corporate management; the AMA's birthing atrocities breaking up Nature's elaborate bonding; schooling's induced madness; the dozens of mercury-laden vaccinations a new child must undergo (the worst aspects of "science-says" superstitions)—on and on, as found in the lion's ever-present cultural guise.

So all such destinies for that new life depend to an indeterminable and varying extent on the mother's emotional state and the care and nurturing of her infant. Her emotional state in the last stage of pregnancy determines not only the successful completion of the fore-brain, but the rudiments of the prefrontal cortex, which begins its growth in that same period shortly before birth. This last-appearing evolutionary structure (the prefrontal cortex) can, however, only be completed after birth, for, were that structure completed in utero, we would never get out of there, so huge will evolution's latest neural system in our head grow. And herein the plot thickens.

The full growth of the prefrontal cortex, which can occur only after birth, takes all of nine months to be completed, and is the most emotionally dependent of all neural systems, radically dependent on the nurturing matrix of mother's arms and breast. Denied this matrix, the prefrontal cortex will be compromised even in its basic cellular growth, and even more so in its function. And the equally critical nature of the development designed to follow—the "toddler period"—is yet again almost totally dependent on the emotional state of mother (or some stable, nurturing surrogate).

What takes place in this last part of the "in-arms" period, overlapping into the toddler period, is the only way all parts of the brain can be integrated into a workable unit. This unit determines the general brain growth for at least the first three to four years of life, and, as Allan Schore points out, heavily influences emotional life, immune system and general health, and intelligence itself throughout life.[3]

Eons were spent developing higher life systems, all of which depended on a well-developed survival system. Even microbes have an "avoidance maneuver" for survival. And the more developed a system, the more developed that survival portion of brain on which such system rests.

The Critical Question

We can say, in a rather dramatic metaphorical way, that at every human conception Nature asks the question: "Will we be able to move toward our higher evolutionary intelligence and possibilities within this newly forming life, or must we defend ourselves again?" The answer in all cases centers on the emotional states involved. Pert's molecules of emotion pack a wallop.

As we shall see in subsequent chapters, such higher evolutionary thrust includes by default the makings of an eventual matrix-for-mind beyond body-brain, the not-to-be forgotten Holy Grail of our evolution, somewhere to go when the lights go out. . . .

Exploring the Prefrontal Cortex

Consider these first three largely independent neural systems—reptilian hind-brain, old-mammalian emotional brain, new mammalian brain with its rudimentary thinking-speaking—each spontaneously responding according to established evolutionary patterns.

At first, lacking any unified, coordinated function, Nature's next undertaking is to coordinate these separate and independent systems into a singular neural network, which, as we will see, has the potential to come under the guidance of heart's intelligence. And for this critical task she begins construction of an even more advanced brain that can coordinate all three into a single, smoothly functioning and superior intelligence. This "great coordinator" and upcoming governor is, of course, that aforementioned prefrontal cortex, evolution's latest achievement. So equipped, the infant and then toddler is ready to "map its new world" from that moment of birth until maturation, when this new brain will reign as "governor" of the whole show, and do remarkable things with it.

Pert's Molecules Again

Throughout this coordinating period the mother's emotional state, determining to a large extent the infant's corresponding state in utero, even more heavily influences the growth of this prefrontal cortex from its beginning through to its completion around the ninth month after birth. This particular genetic group, for this most advanced of all neural structures is the most sensitive to negative hormonal influence of all neural systems. And how Nature's overarching question will be answered, as to which direction this new life will take—toward defense or higher dimensions—will depend on both mother's and infant's emotional states in the womb and beyond.

While birth, and the nurturing that should follow thereafter, is the most critical event of all our life stages, and is generally a major disaster beyond exaggeration, it is also far too large and critical an issue to go into here. The subject has been copiously covered in my previous books and many others, and indeed requires a book to even begin to describe; such descriptions many of us keep offering as prescriptions for sanity, but so far to little or no avail, culture being the power it is. Suffice here to say that abandonment in any of its myriad forms is the greatest and most damaging fear a human infant-child can undergo, and deprivation of nurturing and care is equivalent to major abandonment.

At any rate, such infant sensitivity increases as the neural mass of that prefrontal cortex unfolds and expands after birth, while the same general disaster brought about by birthing lurks behind the scenes thereafter, manifesting in the series of developmental stages that follow.

This neural governor, the prefrontal cortex, will take all of nine months to be sketched in enough for preliminary functions, although its full development and employment will be years down the line. Meanwhile the mother's and her newborn's and eventual infant's emotional states will enter as major factors in the full building and functioning of this "neural governor." If all systems are in a positive emotional state, the structure will be complete enough for preliminary employment by the end of that nine-month period after birth.

At which nine-month milestone Nature responds with another and even more critical structure called the *orbito-frontal loop,* of which importance we cannot overstate. First, consider that from the moment of birth the sensory-motor functions of the infant will have begun that ongoing and lifelong world-mapping as designed (what Piaget called "structures of knowledge"), wherein Nature's multitasking goes into high gear. Two major construction jobs at the same time occupy this first year of life: building the prefrontal cortex, while busily employing the old reptilian brain in its sensory-motor construction of a world-knowledge, which is built only as we interact with that world.

Enter the Orbito-Frontal Loop

At this nine-month point of after-birth growth, all these disparate neural systems must be integrated and brought into a single responsive unit under the command-guidance of that prefrontal cortex, evolved for just this purpose. Without some unification and overall neural guidance we would consist of a mix of blind reptilian instinctive reactions, the unbridled responses of a four-legged creature in the jungle or forest, and/or the unruly, half-baked behavior of an immature, half-brained idiot. Quite a potpourri and not a very attractive prospect. (The reader can probably think of his or her own living examples of each category above, or in such unhappy aggregates of them as found in nearly any politician or even some kin, or—God forbid—perhaps even one's self!)

The prefrontal cortex apparently signals all three units of this budding brain (reptilian, old and new mammalian); these units send branching neural structures (large quantities of neurons, dendrites, axons and such) out to that area where all three brains are in closest proximity with that prefrontal cortex, an area immediately behind and adjacent to the orbits of the eyes. There those neural extensions (axons, dendrites, and the like) from each brain mesh, and begin a large scale neural "looping together," which will link them all into an integrated, functional unit under the direct "governance" of the prefrontal cortex, as Nature designed.

Enter the Fate of the Orbito-Frontal Loop

This meshing, which creates the orbito-frontal loop, is a very large neural organization, almost a small "brain" in its own right, bringing together and linking into a unit of all these evolutionary neural structures. This ties them into functional resonance with the prefrontal cortex, wherein the prefrontal cortex can integrate all into a powerful, unified response.

Our capacity to focus on, consciously attend, and concentrate all energies on a particular event or possibility, lies within this integration. This will in turn integrate that ancient instinctive sensory-motor system (the hind-brain) into service of the higher intelligences of the later evolutionary brains. And then all can be integrated into service of this governing prefrontal cortex and—as we shall see—its connections with the heart. At such developmental point, Nature's perennial question of which way this new life will go seems to be answered in the affirmative.

Construction of this high point of evolution's venture, the

orbito-frontal loop, takes a full three months after completion of the far larger prefrontal cortex (which took nine months itself). Statistically this new loop is completed around the twelfth month after birth, though the actual time frame is quite variable. At this point the infant, under the guidance of its old-mammalian brain, having risen from its reptilian on-the-belly wiggling about into its mammalian crawling about—and now under the guidance of this higher brain—moves into a fully upright, two-legged human stance and becomes one of us.

Wherein that excited toddler, with all its resources under at least some control, is ready to charge out into this brave-new-world and explore every tiny nook and cranny in it. Through taste, touch, feel, smell, listening to, communicating, resonating, and identifying with, the toddler builds those critical "structures of knowledge" on which his future depends.

On entrance of this toddling-ball of energy, curiosity, and ceaseless movement, it looks as though Nature's great question will be answered in the affirmative after all—that child's spirit soaring in high excitement as it reaches out to embrace and absorb its new environment. In my family we referred to this as the "OOOH—AAAH" period of child-enchantment and constant pointing-toward in surprise and delight over its new world unfolding. Which, my friend David Tetrault points out, should be our lifelong state of "living in constant astonishment" at the wonders unfolding.

THREE

THE GREAT CONFLICT

Two Contradictory Commands
Confounding Development

The toddler is driven and guided by two major genetic directives, or instinctual commands, on which our species has depended and a toddler's entire life-drama hinges. First, and foremost, *"Explore the world out there, and any new and unknown event or part of that world you encounter thereafter."* And, second, *"Maintain contact with your caretaker or nurturing-one at all times."* This second commandment is particularly strong when the toddler is outside the family nest, such as in the wilds (an unknown backyard, garden, park, or wherever).

Generally the caretaker with which the toddler must maintain contact is the mother, the one bringing the child into the world, nurturing and protecting it from the beginning, and so a major cornerstone and touchstone of all development to begin with. At times, of course, it is the father, wherein all the same conditions hold.

These two directives of the toddler's actions are ancient and powerful, found in some form in the young of every mammalian species. And these two primal directives—explore, and maintain contact—contain within them the key to full development and the gateway to the whole evolutionary enterprise of our life. With these two powerful commands behind the scenes at every moment of that toddler's experience,

Nature's great question appears: Is it forward in evolution, or retreat into defense?

Every Nine Minutes

Allan Schore's massive twelve-year study focused on the critical orbito-frontal period (see chapter 2), finding in it the most serious and crippling setback in development, and life thereafter. About every nine minutes the average American child's excited exploration of its world is interrupted, cut short, even nullified and prevented, by an equally highly charged negative command, of parent or caretaker: *No! Don't! Don't Touch That! Don't Touch This! Don't Do This! Don't Do That!* Such prohibitions are the foremost directives given and heard every way the child turns. The ubiquitous notice "Keep out of reach of children!" seems to refer to the whole world of the toddler, whether posted or not.

Unless the toddler complies with the caretaker's command, and quickly, the caretaker automatically follows through with some form of reinforcement, punishment, verbal threat, or reprimand: *Do this, Do that—Or else!* And on that *"Or Else"* hangs the gist of the whole affair.

The issue is that parents, almost without exception, had exactly the same happen in their own infancy-childhood. They have themselves had etched into their own brain-body since earliest memory these very same overarching contradictory commands that are an integral part of culture.

This can be summarized in culture's greatest overarching commandment concerning childhood: *The child must learn to mind and obey!*—with its myriad "Or Else" injunctions. This commandment is etched into the minds and memories of every enculturated person, and emerges as the most powerful of directives on becoming a parent. Neither parent nor child has much to say about the issue. Darwin's stipulation holds: any practice repeated long enough becomes habit; any habit repeated long enough locks into our genes, and surely thus in memory.

So on the child's obedience to this cultural ultimatum, the parent's own social image or identity hangs. Success-as-parent in the public eye, even social acceptance in general, seems to hang in the balance here. Parents feel judged by their society according to their child's behavior, as they and their parents were: Does the child meet the social-cultural patterns of behavior or not? Since we have all undergone this "upbringing" and respond in the same way, common sense and personal integrity, as well as conditioned reflex, demand our compliance with such commonly shared, unquestioned, and "common sense" belief. And as it is surely common to all, on this common fatal flaw we all go down, generation by generation.

A toddler's newly forming neural system functions according to these two formative commands: explore and maintain contact, almost from the beginning of its life. So long as the child is safely in crib, playpen or such, all smiles are on him. The moment he gets up on his hind legs to move out to explore as nature directs, all eyes are on his every move in judgment of those moves (and always "for his own good").[1]

This constant scrutiny, with its prohibitions and interferences, creates a contradictory situation for toddler: caretaker on the one hand, and inner directive for exploration on the other, both demanding attention-energy, yet each virtually cancelling out the other. (Years ago I heard it claimed by animal trainers that you can drive a dog mad by training it to follow two directly opposite commands and then issuing both at the same time.)

Struck by two major evolutionary signals that completely contradict each other, the toddler is driven to explore on the one hand, while on the other is commanded not to. Meanwhile a further contradiction arises when that care-taking, protective-nurturing person becomes adversarial and *threatening*. Where, then, does that new life turn? Abandonment threatens from every quarter.

In this bewildering maze, the child has no choice but to automatically try to maintain contact with his caretaker, and yet move on in the other directives he also must follow. Splitting his attention between Nature's inner directive for exploration, and the outer social-cultural demand *not to* (issuing from his safe-space protector herself), the toddler complies as best such a divided system can. His emotional-sensory motor system moves those muscles and limbs on to explore, while his relational-higher brain connections move him to maintain contact with that safe-space, care-taking, nurturing one.

As stated by developmental studies over and over for years, separation, isolation from, or abandonment by the caretaker is the greatest fear mammalian infants—or children in general—can experience, this being critically so in humans. So either way he goes, toddler's confused response is not a willful decision, but a directive his inherited and ancient instinctual reactions have made from the earliest beginnings of mammalian life—now leading to paradox.

The power of this negative command concerning exploration, and its parallel danger of separation, lies in the simple fact that an infant-child abandoned or cut off from its caretaker, was, throughout mammalian history, generally *saber tooth's lunch,* in one form or another. So this ancient instinct, involving that powerful amygdala and its links in both reptilian and old-mammalian brains, has built into all mammalian systems one of the strongest of all evolutionary imperatives, setting up a deadly roadblock to a response we humans are driven to follow. This apparently harmless and absolutely common-sense negative command— No! Don't!—which we parents toss out so casually, rides on the wings of death for the wholeness of the child's self—and no small consequence to our own self down through the years.

For most people to accept, much less believe, the implications of this scenario, seems near impossible. My own efforts, and those of many wiser than me, have met a blank wall on this issue for decades. All of our own neural systems and structures of knowledge have had this very threatening survival-directive of minding-obeying built into every cell, as a security measure itself. And security is no easily ignored factor. Further, overriding even this automatic reaction is the equally real issue of our own public image-as-parent in our society and culture's judgmental eye. Recognizing the fact that our obedience to our parents brought and brings a heavy price in us seems outright nonsense to most people who "know right from wrong and have some common sense—after all." Such a notion concerning child obedience arouses a thousand "what-if" qualifying scenarios in our minds.

The only alternative is the corollary fact that the child imprints from parent-model's behavior every bit as much as from these parental negative commands. But following through on this would demand serious and rigorous behavioral and attitudinal changes in a parent! This makes the behavioral matter even more difficult for the parent to comprehend, and is generally rejected outright. It is "the child's behavior, after all," and the parent's duty to modify it according to "common sense." Around and around, the wheel grinds on.[2]

Some Mechanics of Candace Pert's Hormones

At any threatening or negative command from the caretaker, the child's ancient defense system releases into its young brain a hormonal burst of adrenal-cortisol that brings a powerful "alert response," redirecting all neural responses to shift to attending that primal defensive system, while putting everything else on hold until this warning-alert is tended.

The peripheral nervous system goes into survival mode, and energy from the higher cortical systems is redirected to the energizing of the lower survival instincts, prompting the toddler to reconnect with that caretaker and re-establish its safe space, even as it goes on guard. Generally the toddler makes such a re-connection by trying to make visual contact with the parent's eyes.

At the same time, paradoxically, the equally powerful ancient directive for exploration is renewed and re-invigorated, just by the burst of adrenals if nothing else, and the two drives try to function at the same time in diametrically opposite directions. Many a time we have heard some frustrated mother or father report, "And that little devil looked me square in the eyes and did exactly what I told him or her *not to do*!"

The sympathetic nervous system, with its defensive release of negative hormones and tightening of body-armor, continues, well after such a restrictive episode has passed, and before the parasympathetic system can re-establish unity and calm. (As the sympathetic nervous system is designed to turn the alert on throughout the body, the *parasympathetic* is designed to turn that alert off, slow things down and re-establish coherence, relationships, balance, and harmony.)

According to widespread research gathered and published by HeartMath,[3] such rebalancing of hormonal and neural systems after a negative experience can take up to several hours before balance is restored, depending on the nature or severity of the disruptive episode.

Every Nine Minutes (Again)

Yet, Allan Schore's research claims the average American toddler undergoes such split of directives, on average, *every nine minutes,* leaving almost no time between such negative events for parasympathetic rebalancing. And even if cortisol release might cease, its residual effects of tension and fear linger on, even as the child, under his other "high-command," generally overrides caution and resumes his exploration as best he can.

Most parental commands of this *No-Don't* order have to be issued time and again, since they go against our genetic encoding on so many levels. And sooner or later such demands will be backed up, in frustration, by physical means: a swat, shoulder-shake, or isolation. Or, more frequently, increased and ever-more-harsh verbal threats. "I'll beat the blood out of you," was the favorite of a parent in our neighborhood with a particularly loud voice and "obstinate" child.

Even more damaging is the threat of withholding or withdrawing love or nurturing, or actual casting out physically: standing in the corner or broom closet in shame. "Do this or that and mummy-daddy won't love you any more, nor even *want* or care for you," is how these commands are too often spelled out in the child's mind.

Eventually the child learns not to follow its impulse to explore until he checks it out with the parent or caretaker, and even then will keep one wary eye on the parent (or authority) and the other on the venture undertaken—a splitting of attention that compromises a child's natural inquisitiveness and spontaneous response, as well as its ability to attend single-pointedly as attention demands.

Hesitation and self-doubt generally result, a doubt that etches into that growing character as a critical flaw. Eventually, true attention—where all body-systems coordinate to focus on a single issue, the job of the prefrontal cortex—will be fragmented, and some degree of "attention-deficit" disorder displayed, as is rampant today.

Re-Routing the Orbito-Frontal Loop

The most serious result of all these split directives and errors we make with children takes place within the very neural organization of the orbito-frontal loop itself. By about the twentieth month, some eight months after the great conflict begins, actual and dramatic changes in that orbito-frontal loop's neural structure, as well as its operations, will have become readily evident (as Schore makes clear). This is a simple neurological fact never considered before Schore's work, and the most serious of all single aspects of a child's life. To overstate the seriousness of this fact is hard to do; yet, for it to be accepted presents an obstacle even more difficult. Of all child-parent interactions, this resistance to change in a parent's belief is the most difficult for a parent to recognize.

Schore's research further demonstrates that as a result of these enforced and ongoing behavior modifications, the very neural "vertical and translateral" links within that orbito-frontal loop, connecting all four systems and generally established by the twelfth month, are dissembled and re-assembled in a number of disastrous ways. The "vertical" links between the high-level prefrontal cortex and lowest sensory motor defensive system are reduced in varying ways and to a varying extent, while the "translateral" links between the emotional brain and that defensive system are strengthened.

Those vertical links are where all systems are designed to be lifted up from their lowest level into those higher orders of functioning in the prefrontal cortex. The translateral links are between the emotional brain and defensive hind-brain, which are enlarged and strengthened, even as the higher vertical links with the prefrontals are compromised, most are re-routed into translateral connections. When Nature feels forced to reinforce her "translateral defenses" for protection, at the expense of vertical movements toward our higher intelligence, once again Nature's great question has been asked, and once again answered in the negative.

How, in the face of such confusion and paradox, could a young self move on into the higher realms of intelligence when it has no organized, cohesive structure *for* such movement, or safe space from which to operate? How else, but that his basic intuitive instinct be impelled to increase defenses and retreat yet again from movements into unknown higher realms?

Research shows that by about age three, these early events bringing modifications in the orbito-frontal loop in those first three years will have myelinated, which makes them permanent. Myelin is a fatty coating that forms on axons making up a neural field, such fields as make up the orbito-frontal loop. Myelination makes such fields impervious to the hormonal actions of the parasympathetic system's periodic house-cleaning of unwanted or useless memory-information, such as those left over from some sympathetic nervous system's alerting fear or threat of abandonment—real or imaginary—in earlier times.

Making permanent the fear-alert patterns suffered in the first three years affects our lifelong mental-physical-hormonal responses, the sixty-year-old just as strongly as the six-year-old, since these are permanent shifts in then-permanent neural structures, which function on levels below our awareness. Those correctives, reprimands, or threats of abandonment, which so disoriented us as toddlers, will affect our sympathetic system thereafter, in any number of different forms, generally below the threshold of our awareness.

From that point of myelination forward, any incoming signals that are in any way resonant with those original negative events, which split our attention and caused distress way-back-when, will trigger that same response again and again, in ever so many different, subtle ways and resulting in ever so many different, subtle effects. Most of these effects will be below our awareness, though showing up readily enough in a myriad of not-so-subtle signs as nervousness, irritability, fear, chronic anger, high blood pressure, cancer, or other unhappy afflictions of our modern enlightened society.

And underneath all parent-child relations will be that parent's own fear, triggered by associative actions with their child's behavior, of the social approval or disapproval of them as persons—as well as parents—regarding their child's own behavior. *Have you taught your child to mind?*

Some recent claims call our attention to the brain's plasticity, which seems to qualify earlier convictions of neural systems being impervious to re-formation once formed. This plasticity of the brain is not an over-all automatic response, however, and certainly not a handy cure-all, but at best only a possibility that can, with care and patience, perhaps help bring critical remediation to some damaged systems. But to take plasticity for granted as a loophole corrective for faulty development, rather than taking care to bring about a balanced system in the first place, is risky, and at present is not paying off. Prevention and care are always the first imperative.

In a final look at the fate of the orbito-frontal loop, we find that variations of such re-routing and shifting as take place in the toddler also take place to varying extent and in various ways at each major shift of development thereafter. They appear in the four- to six-year-old shift that will end the period of the dreaming child and bring in the operational logical stage, itself ending with a major neural pruning around age twelve, as well as in the tremendous shift into adolescence that pretty well scrambles parts of the adolescent brain for years. And on it goes.

Most of these neural shifts will reflect the effects of this early permanent re-alignment of the orbito-frontal loop in the toddler, with its parallel realignments and shifts in the amygdala and other critical memory and defense structures.

One final observation must be made of that toddler-child, driven instinctively to investigate its world to at least a minimal survival level, while also instinctively trying to modify its behavior according to parental-cultural demands (also for survival). This child develops an actual division of his previously unified self-system—his innate identity and integrity. The self-as-brain finally orients to the lowest physical survival patterns, while the complementary self-as-heart orients as best it can to the higher mental-imaginative-creative-spiritual and social imperatives inherent within the prefrontal-heart connection. This brings a fundamental split of self in what evolution designed as a simple division of labor in her two greatest and most fragile creative systems—heart and brain. Instead of a unified self, we end with a split self brought about through enculturation and behavior modification—a self now at war with itself.

Judged Guilty Until Proven Innocent

From abandonment or the fear of it, that new life is compromised and brought into nothing less than servitude to the more powerful social-cultural pressures to which that person has had to adapt. The overriding judgment of being unworthy until proved otherwise, and the underlying feeling of guilt or failure until proven innocent, result from those chronic negative directives of culture and the self-system's struggle for integrity. These cultural imperatives have created a double bind wherein self is wrong any way it turns, and they will plague that person's every move. This underlying guilt and these feelings of failure will drive him or her to "prove themselves" to that accusing world, and so rid himself of the accusation of unworthiness, at all costs. At all costs, because this is essentially a matter of survival of self, of one's basic capacity to orient in the world in a coherent manner. And the cost—which is borne equally by that world, social body, and self—is of no small accounting.

Which brings this little survey full circle. Examining our "human condition" has opened us into a labyrinth in which we have been lost for centuries, while blindly seeking a way out. Yet those very systems in question were designed—as we well intuit in our deepest knowing—as a way of transcendence into ever higher realms needing no way out, but simply a *moving into*.

FOUR

CULTURAL DEFAULT AND INTENTIONAL EVOLUTION

A Worm in the Heart of the Rose

A most basic, sacrosanct, fundamental fact of our life today, an issue no more questionable to a sane person or society than would be to question gravity itself, is the axiom that *ambition,* the burning desire to "get ahead," to be somebody, to amount to something, is the most natural and desirable of all traits a person can have, particularly our young.

Since such a drive is culturally induced and an automatic reflex in us, for me to declare, as I do, that this so-called ambition, honored as the greatest of values and virtues, is a demonic cancer eating at the heart of life, Earth, and the human spirit, seems a ridiculous and dramatic overstatement. Induced like a disease and sustained at every hand, this desire for cultural purging hidden within, is a need that drives the corporate executive to batten on his brother's blood, at home and abroad, to further that executive's fortunes—*at all cost.* Such fortune must constantly grow, more and more, since the need involved is a chimera, a fantasy, nothing other than a form of paranoia. And the replications of this drive, emulated in society at large, are legion and costly.

To prove our worthiness or justify our existence, we enculturated minds move, regardless of price, for the success of any goal, which we think might rid us of this inner conviction of failure (which adds to our underlying need for a safe space and its security). While the price paid is generally high—for self, spirit, society, and world—it enriches and enhances culture as a field-effect, as well as the powers behind its judgment-system, those few who "make it to the top."

Emotional Fields and the Inversion of Evolution

The emotional index presented in the last chapter indicates, by extrapolation and extension, that emotional turmoil in a society or environment might spur production of infants with bigger hind-brains, bodies with more muscular mass, and so on; and that a placid, calm environment would have mothers produce infants with bigger frontal brains and smaller bodies, which seems evolution's intent (a note equally pleasing with me).

Candace Pert's work shows that "molecules of emotion," positive or negative, will be incorporated into a genetic pattern as naturally as any of a number of influences, an issue that weighs into this book's discussion of the two-way traffic between our genes and environment. And surely, Pert's work throws light on our mess here in the twenty-first century.

During World War II—wherein our species violently murdered, by every hideous, painful, and horrible device, well over a hundred million of itself—surely survival reflexes were unleashed at a maximum peak for at least six unbroken years, with additional overlap of years on either end. These were times of extreme anxiety both at home and abroad, as well as overseas with the troops.

World War II was followed almost immediately by smaller wars, leading to ever-larger stockpiles of ever-greater forms of mutual annihilation, involving more and more nations until leaving no one out of the deadly equation, and surely "no place to hide."

WW II veterans back from overseas slaughterhouses were now busy building cold war bomb shelters in their backyards to try and protect their families, the spawning of which families had kept those semi–shell-shocked boys busy since they first came back. Mounting waves of crises rippled across the spreading worldwide population as well, busy spawning and rapidly expanding their populations too, while mounting ever-larger waves of negative emotional force. Setting up such powerful reciprocal interactions directly or indirectly involved virtually all life on Earth—plant and animal—crises hardly resolved half a century later here in 2009 as I write.

Judging from the brief article of 1998 referred to in chapter 2, and the ongoing research following, one could speculate that the overall emotional state of the ever-increasing number of women giving birth was at least tinged by sheer proximity with this all-pervasive stress and concern flooding most nations. And, in the postwar population explosion beyond record, a majority of infants were born with enlarged hind-brains, somewhat diminished fore-brains, and correspondingly larger skeletal frames and heavier muscular bodies, increasingly tilting the balance of nature toward the negative, defensive end.

Indeed, as history shows, the population of "baby boomers" brought a generation that stood some six to eight inches *taller* than its parents (to which I can personally attest, judging by my five looming over me). This extraordinary increase of stature brought, of necessity, large expansion and growth of all physical facilities, infrastructures, food production and consumption, even beds and furniture and on and on.

The size of one's offspring became a critical issue to parents' pride. Bigger was better and more beautiful in all situations, parents ever so proud of their children towering above them, those children looking down on their parents. Meanwhile, the runt of any litter (never ask me who) rather lost out, tending to be unaggressive, less ambitious (a failure, if not sin, in this new world), but of necessity less caught up in the "race to the top" and "place in the sun." In which race some of the aggressive average were always winners, though most, of course, were losers.

Male children gauged each other according to size, the larger and more aggressive the more popular and in demand, wherein growth hormones became an issue, synthetics eventually appearing. The mental set of this rising population reflected the increased hind-brain survival-hormonal influence growing at every hand, an influence touching most aspects of our lives, including sexual maturation—an issue heretofore considered "locked into" our genetic system and inviolable.

An odd statistical item in the genetics of reproduction can be attested by any astute gardener. If you want to get the earliest tomato in your neighborhood, try deliberately slicing down into a young tomato plant's root system with your shovel, severing some of the roots. That plant will tend to produce a single bloom and fruit well ahead of the other, undamaged plants. In the same way, animals will tend to reproduce much earlier if general conditions are threatening. A rule of thumb seems to hold that early damage or ongoing anxiety-fear in a growing creature will incline that creature toward reproducing sooner than average, as though Nature says, "You might not be here long, so reproduce your kind while you can," which, we might say, upholds Nature's quota.

Another oddity seldom addressed was the striking change in menarche (the beginnings of menstruation). In the general post-war period, menarche shifted from a traditional mid-teen average age of fifteen years, down to age twelve (and even earlier) by the 1980s. This mid-teen average had held historically, occurring a bit earlier toward the equator, later toward the poles (in Scandinavian girls later, Mediterranean earlier).

By the late 1980s the Child Development people at the U.S. Department of Health, Education, and Welfare (HEW) reported that forcible rape of girls by boys under ten years of age had become an issue, although originally a statistical oddity at best. Male sexual capacities had historically appeared only in mid-teen years along with menarche. One could speculate that the species in general had speeded up the reproduction age for the same reason as that threatened tomato plant.

Other Age-Related
Social Shifts

Some three generations back, our backyard sandpile play as small children expanded in later childhood into such games as "capture the flag," sandlot pass-and-tag football, baseball, and so on. These were carried out entirely on our own and were the source of serious, engrossing *pleasure,* sheer fun, and, looking back at this decades later, an excellent social-cooperative "training" we had never heard of, and couldn't have cared less about had it been so named. Winning or losing, however, played almost no part in our play of that period, since "sides" in backyard ball games changed continually as parents called individual team-members home, bringing a general re-shuffling to maintain balance of sides (a serious issue). No side "won the day," since sides were not fixed, and scoring, beyond three strikes and you're out, never entered our heads. We played ball to have our chance at bat—the strangely glorious feeling when bat connects with ball and that ever-so-satisfying sound found nowhere else! Or that rare pleasure at catching that high ball, or pitching a true curve ball. We played passionately for the sheer sake of play, even as we were building a strong social sense of cooperation and "fair play."

Over time, play gave way to, or rather was taken away and given to, first, organized playgrounds with adult monitors and supervisors (lest someone get hurt and lawsuits result), and later, organized sports. We had never called or even thought of our backyard or vacant-lot ball games as "sport"—just play! But now this became sports under the supervision and jurisdiction of adults. Taking over, these adults turned play into adversarial contests ruled over by coaches and spurred on by parents and spectators, the young participants caught up in bewildering and anxious attempts to satisfy those supervising and goading adults, suffering guilt when failing to do so.

Play Dying on the Vine

Such organization of "sports" moved into childhood earlier and earlier, spurred on not by children, but by parents and "coaches." Little League, a most insidious attack undermining childhood, took over, the small fry decked out in bright colors advertising various products. "Coaches" became a principal figure in those young lives, coaching an immediate target of adult male ego-image. The fierceness of "the coach's" strictness and harsh demand-commands became a tradition and mark of adult manliness, and a toughness that children eventually tried to emulate. The rough stimulus and overall threat of shaming, which the coaches wrought, gave a complete cultural model-image of The Coach, with his shouting, threatening, berating, bullying voice and profanity heard over the land.

Gladiatorial types of sports grew in the place of play, amid waves of near-frenzied adult excitements over adult college and professional teams struggling for dominance in the arenas cropping up everywhere to accommodate the phenomenon.

A survey in the late 1960s of high-school students' definitions of manliness and masculinity revealed that the ability to deliver harm or pain to another without flinching or remorse was the principle mark of a manly, adult character. Tough-man stances, clenched jaw and impassive facial expressions, along with a "fiery temper," reflected manliness; smiling pleasantry was an indication of lack of character or strength of conviction, perhaps even a "questionable sexual orientation." Romanticized mythical images of such toughness and masculinity were slowly emulated by the entire society.

All this encouraged and even glorified the ever larger and more combative-competitive young people these communities helped produce. The aggressive drive to "get ahead" or outshine others—thus escaping the guilt of failure—spilled over into the schools, marketplace, town hall, Wall Street, Washington, London, Brussels—on and on.

The advent of television found all this a near-perfect soil for its

own explosive growth, and brought increasing waves of rabid attention to such combative events, captivating the majority of the populace, promoting sponsors and selling products, spurring more and more
consumption as well as competition. A whirlwind of ever-increasing
proportion reflected in multiple layers of various neuroses and phobias,
while irrational violent behavior became near epidemic—the general
mind-set, expectation and direction of the global populace.

Reciprocal feedback, creating ever-wider fields of negativity, resulted
in a worldwide tangle of raw emotion, maintaining a state of tension
and intensity in all aspects of life. Conditioning to such overloads of
intensity (to which television is a heavy contributor) brings—should
such intensity be withdrawn—the equivalent of *sensory deprivation* to
both young people and older ones. Anything less than a television level
of adrenal-stirring excitement tends to become disquieting and unset-
tling. Camping, fishing, and hiking, passionate pursuits in my day, rank
at the bottom of the list to teenagers, "pre-teens," and older children
today.

In my section of Virginia, the deer population has grown far
beyond previous levels because, among many factors, in recent years
deer hunting has diminished markedly. The former hunters now sit riv-
eted to their TVs watching the game, sipping beer, eating munchies,
and getting fat, while the deer sport unmolested through our gardens
having their own munchies and getting fat. TV sports with their mass
audiences and passionate team loyalties bring adrenaline rushes to all
onlookers, stadium or screen, supplanting, in a form of proxy, the keen
adrenaline rushes we experienced in earlier days through play, hunting,
hiking, backyard sports, and such.

More! More!

Life without television seems lackluster to most citizens—boring, of no
interest. Further, such antiquated habits as we once enjoyed contributed
nothing to the frenzied demand for growth. Massively sponsored by

television, and expressed by nearly every aspect of life since World War II, growth, stimulated at every hand, constitutes the very breath of our nation's economy, politics, education, and mind-set. Steady, viable maintenance, requiring the balance of all Nature, as found in the preceding century when 95 percent of people lived on farms, has given way to this dominant demand for growth—at all cost. This obsession became an outright survival issue to the entire fiber of our national thinking, even as nonviable and unchecked growth was choking our living environment and planet itself.

"More! More! Is the cry of a mistaken soul," claimed William Blake, "less than All cannot satisfy Man." Blake's general All referred to the evolutionary goal of our species' longing of Spirit, in its various guises. Demonically inverted, All is the cry of a nurture-starved, fear-based humankind, to whom possession of all the goods in the world could never satisfy, though most of us might die trying. Since in size, influence, wealth, dominance, and popularity, *grow or perish* is the byword today, if we are not growing our economy, our pundits warn, we are surely dying. And the same yardstick seems to apply to life in general.

The eighty-year-old, shaky and trembling on the edge of his grave, is elated when the stock market goes up, depressed when it goes down. While the vastly overpopulated Earth, decimated and stripped by the voracity for growth, gives but the grounds for ever-more rapacious growth, the social body become cannibalistic, feeding on itself.

Reconsidering the Heart

As I pointed out in my book *Evolution's End,* heart is the source of intelligence, while brain-mind the source of intellect—which was evolution's intent in separating self into two reciprocal, interactive functions. Rather like right and left hands, intellect and intelligence are a "strange loop" phenomenon, wherein each gives rise to and is critically dependent on the other, while equally requiring their independent functions.

Faced with the plethora of potential in our world, an enculturated mind, laboring under its weight of induced guilt-failure, triggers its selective attention to attend any potential, which mind thinks might lead to success, fame and fortune, self-justification, pardon from cultural-guilt—and possibly even its missing "safe space." Trying to find a "safe space" drives most enculturated people, suffering as we do from early failure of bonding and nurturing. A safe space finds its proxy, or counterfeit, in material goods and possessions, while the only real safe space for self and Spirit, of course, lies within.

Concerning such rich potential as our world offers our selective attention, in regard to some possibility it intuits or senses, intellect asks only, *"Is it possible?"* In the same situation, the heart, with its intelligence, asks, *"Is it appropriate?"*

An enculturated intellect—which includes virtually all of us—is driven to justify its existence and prove itself worthy in the eyes of that culture, through any possibility afforded, appropriate or not. Stripped of self-worth through enculturation, we are accused of worthlessness until *proven worthy* in culture's eyes. Thus, we harbor a hidden feeling of guilt for not measuring up, which failure is, in effect, a "cultural sin" lurking within us. This covert notion of failure drives us to do anything that might "measure up" to culture's standard and so be forgiven, in effect, by that culture, which is rather like seeking absolution of sins in the middle of a drinking spree in a whorehouse.

Heart's intelligence, on the other hand, moves only for *well-being* of self, body, spirit and world. And no greater contrast can be found than that between an enculturated intellect and the open intelligence of heart. Intellect and intelligence are so close, inter-related, and interdependent when functioning as an intact strange loop, but so far apart when split by culture. Through enculturation, heart is not only never heeded, but almost never heard, the lines of communication between heart and brain-mind having been compromised, or near eliminated by enculturation—and this at every level: biological, psychological, spiritual. (We do well, of course, to never underestimate the brilliance of this human intellect, nor apologize for it, even in its most base and destructive forms and actions, wherein we simply represent a split of self and God gone wrong.)

Appropriateness, as granted by heart, brings coherence, balance, and an open-ended present with a future flowing into it, free of consequence or past influence. "Behold, I make all things new," that great being explained, a newness that does not, and cannot, include items of the outdated world-self view of the person actually opening to heart.

Unbonded and unnurtured, mind's connections with heart are not only compromised. Our resulting intellect, lacking heart's guiding intelligence, becomes our nemesis; it is slowly destroying us, generation by generation, the tempo speeding up now with technology at culture's disposal. The smarter we get, the more dangerous we are to ourselves.

Since every fiber in the fabric of our current mind-set religiously believes in and relies on our rational, scientifically sanctioned intellect, any alternate actions of mind outside this cultural loop strike us as irrational nonsense, although an actual "way out" for us lies within such alternate ways. Yet, as suggested in chapter 3, we should be cautious of the "way-out" mentality, which can easily slip into techno-mechanical "fixes," landing us right back where we started. What is called for, perhaps, is a "moving into," which allows the future, as it were, to penetrate the closed cultural looping.

Navigating the Unknown via the Heart

Culture counters any understanding of opening to the heart by the cultural counterfeit of heart intelligence as sentiment, as sweetness and light, sympathy and projected self-pity: "Aw, come on now, have a heart." True opening of the heart, however, requires suspending, letting go of, not only intellect—its rationality and comprehension of self-and-world—but also our ever-present and passionate drive for survival.

In this sense such opening is resonant with discovering what in previous books I have referred to as *unconflicted behavior*.[1] This is a state of mind in which one has thrown away self and its survival concerns, wherein the ordinary "ontological constructs" are no longer binding on one (that is, for instance, where fire doesn't necessarily have to burn, nor gravity hold).

Opening of the heart goes one step further, however. It involves not just a suspension of concern over survival of body-brain (as in unconflicted behavior), but just as critically, suspension of concern for our image of self in the public eye—identity of self as rational, responsible, and respectable. In suspending our survival drive we open to new possibilities, but in opening to the heart we abandon all possibility and identity as *known,* giving our self-as-mind to that self-as-heart—and no longer claiming jurisdiction over either. For our identity no longer tries or needs to be in charge.

This is not an abandonment of personal responsibility, which is a cultural criterion many people would welcome, but rather, far more extreme. This is the abandon found in George Fox and Jesus, throwing away their selves, including that noble and respectable self-responsibility admired by culture. Having thrown away even this last vestige of cultural approval, one then finds one's self in a broader and unconstricted frame of reference altogether, the only state through which one can become an instrument of Spirit. (I speak of this with conviction, having drawn back time and again ad nauseam from the brink of that unknowing, even as sensing its freedom, in a peculiar form of interior terror that is my undoing. That is, I "chicken out.")

Here, I recall Suzanne Langer's observation many decades ago: "Our greatest fear is of a collapse into chaos, should our ideation fail us." This ideation is the cultural mind-set—our reality-picture and our presence in it—which rules our thoughts, beliefs and living presence from birth. It makes for a double bind that only Self-as-Heart can go beyond.

FIVE

ANCIENT HEART-KNOWLEDGE IN THE TWENTY-FIRST CENTURY

At our first meeting, Baba Muktananda immediately launched into an explanation of the functions of the heart and its role as the center of a complex cosmology-ontology, as outlined a thousand years earlier by one Abhinavagupta. (An ontology is an explanation of how our reality-experience forms.) This meditating Shaivite sage in Kashmir, India, handed down the theories, descriptions, and practices giving the foundation of Muktananda's worldview and meditation practice here, ten centuries later.

"There is only one heart," Baba explained, "the one beating in my chest is essentially the same as the one in you." Yet, as his theory revealed, the heart has an individual aspect and takes on characteristics of the ego developed in our head. (The word *ego* is Latin for "I," which I interpret as *me,* and is not entirely a bad-boy in my lexicon, as in some current usage.) "Billions of egos in heads up there, only one heart," was Baba's comment, in outlining this dual yet singular cosmology he called Kashmir Shaivism.

He further claimed that in the center, or "cave" of the heart, there is a point from which the entire universe arises and radiates outwardly.

This locus or central point of reference was called Shiva, and was considered a nonmoving male force, while the radiations outward from it, termed the "wave-forms of Shiva," were called Shakti, a feminine Sanskrit name. These wave-forms contain the potential of all creation, his theory explained, out of which this Shakti creates and gives birth to the universe. (Thus, the feminine name Shakti, birthing having always been considered a ladies' prerogative. Wise men were considered the noninterfering, protective witnesses of such, a wisdom lost long ago in the chaos of the muddled-male-meddling which took over.)

Muktananda made a wave of his hand when describing Shakti's performance, claiming she dances about this nonmoving Shiva in a spiraling gyre that embraces within it the potentials of all conceivable universes. From this plethora, instant by instant, Shakti brings forth a world for Shiva to witness. Without Shakti, he pointed out, there is no Shiva, and vice versa, each giving rise to the other in what I now know as a classic mutual-mirroring, or "strange loop."

Meditate on the heart long enough, Baba said, and you will begin to sense the presence of these wave-forms enveloping you like a warm cocoon of love and power. And indeed, this slowly became a felt presence for me, while twice in my life I directly experienced, in ordinary wake-state awareness, this embracing "cocoon" of love and power. This is a phenomenon that must be directly experienced, I suppose, to be believed or understood. Once experienced, however, understanding is incidental and the reality of this "heart intelligence" can never be doubted—understood or not. One simply knows that "it" is always "there," heeded or not.

"The scientific world," he went on, "will eventually give you all the information you need to explain this Shiva-Shakti creation," in ways available and reasonable to contemporary audiences. This proposal I found the most improbable of all his explanatory prognoses, yet it proved to be the case as time went on, and in 1981, I gave my first mind-heart talk at the University of Colorado in Boulder, followed by hundreds over the years and over the globe.

Baba Muktananda and the Out-of-Body Experience

In the 1970s, through a chain of classic "paranormal" or "psychic" events, I met one Muktananda, an Indian meditation master or "guru." This quite esoteric and unusual (to me) meeting with an Indian sage with his ancient legacy had been preceded by taking one of Robert Monroe's equally esoteric but safely modern "out of body" programs, and the nonordinary adventures therein. Monroe had experienced this out-of-body event many times, exploring, researching, analyzing, and eventually designing a way to induce something similar in others. The institute Monroe established at that time, and the mind-expanding training it fostered, has since grown to serious dimensions. My out-of-body experience induced by the Monroe program had preceded my subsequent meeting and joining forces with Muktananda, a far more esoteric venture, which absorbed me for a number of years, before I got back to the out-of-body Monroe issue.

Through a series of rather odd random-chance events in his early forties, Monroe experienced going "out of his body," as the phenomenon is called. Fascinated with such a state, he followed through with many experiments and repetitions of the experience until he was finally stuck with the procedure, wherein he couldn't really sleep normally any more, since at each attempt he simply went out-of-body again. (Muktananda often hinted at such pasttime at night, saying we wasted those sleep hours that were meant for other and greater adventures.)

At any rate Monroe made a thorough analysis of this out-of-body affair, and mapped out the procedures he had used in an analytical and logical fashion, so others might do the same should they be interested, and the Institute carrying those ventures on flourishes today. Thousands have taken this training and their accounts are often incredible.

My brief encounter with this "Monroe" venture led, by a roundabout route, to my meeting with Muktananda, whose detailed description of this Shaivite theory of creation struck me as the wildest nonsense at the time. Since I had been led to this meeting with Muktananda by what would surely appear to others as equally wild or even wilder nonsense, I heard him out, to my great gain and benefit.

I had retreated from the social-cultural world of book writing and lecturing some three years before this meeting with Muktananda. Gardening and meditating happily in a remote section of the Blue Ridge mountains, no roads in to my place, no power lines or telephones within two or more miles, no newspapers or radio to relate calamities I didn't need to hear about anyway. I had, accompanied by my wife and daughter, seriously and genuinely renounced the world of folly and avoided it successfully for three wonderful years. (Besides immediate family members, only my publishers had my address, that I might receive any stray royalty checks that might come my way—and those publishers honored my request for privacy, though too seldom sent a royalty check.)

Then came the series of paranormal events (recounted elsewhere) leading to my leaving my haven-hideout to meet with this Baba Muktananda. This meeting, as it turned out, ended with my family and me living for ten winters and one summer on the other side of the world, in Baba's meditation ashram in Ganeshpuri, India. This world within itself unfolded as an ongoing inner adventure of which I would have been embarrassed to speak about in my earlier academic days, but learned to cherish and speak about continually in many hundreds of talks given over the years on behalf of Muktananda's "meditation revolution."

Muktananda had strongly urged me to resume writing and lecturing about children and development, the "worldly task" I had abandoned to that chaotic world I had rejected. "In this Yoga," Muktananda said, "we don't meditate in caves, but right out on Main Street, getting our nose bloodied with everyone else. And," he added, "should you want to stick with me, you will have to go back to writing and lecturing."

As it turned out, my experiences with Baba and in his ashram were far too powerful and rewarding to leave, and I stuck with him and his far-flung global meditation group for twelve wonderfully rich years, even after his leaving in physical form (wherein I simply switched to his successor, Gurumayi, who carried on the tradition). Dutifully going back to writing and lecture travels (which grew more and more frequent), I hung out in this Siddha ashram in India at every opportunity.

I gave my first mind-heart presentation in 1981 at the University of Colorado in Boulder, the continually unfolding scientific discoveries I purloined ad lib from a variety of sources displayed in full. In subsequent years I gave over 1,500 such presentations for Siddha Yoga (Muktananda's international organization) in some twelve countries. Meanwhile I turned out more books while travel-lecturing, with a total since 1971 of nine books and near 2,000 talks to date, which is to say, I was energetic and busy back then.

A major issue of our heart centers around a "frequency field" emanating from it. In the 1970s Karl Pribram, now professor-emeritus from Stanford University, proposed that the brain draws its information needed to create our world-experience from a "frequency spectrum or realm" not in time-space, but, in effect, giving rise to time-space in our experience, all of which proved to arise and radiate from our heart. This was an absurd-sounding notion back then, but fit Baba's Shaivite theory of heart-brain and was but one of many contributions adding, over the years, to my understanding of the heart's role in mind's experience.

Contemporary Echoes of Heart-Knowledge

In the earlier part of the twentieth century, the Austrian philosopher-scientist Rudolf Steiner detailed at length how the universe arises from a point at heart's center, in a frequency-form that vibrates out-and-in untold myriads of times a second (as do many micro- or "subatomic" events such as neutrinos or microtubules). Steiner claimed that the whole universe was contained in those radiations from our heart, and made the prognosis that the greatest discovery of twentieth-century science would be that the heart is not a pump, but profoundly more. Further, he claimed that our species' greatest challenge, following our understanding of the actual heart-function, would be to allow the heart to teach us a new way of thinking. Through this new way of thinking, the heart would find its own next step in evolution.

Only in recent years, in looking back, did I realize how fully Steiner had spelled out in clear detail the universe enfolding into that singular point in the heart, to then unfold and expand back out into its full panorama in a kind of oscillatory or vibratory manner. One could have assumed Abhinavagupta had reared his prophetic head somewhere in Steiner's background. More likely, the two simply shared the same subtle, collective memory-fields, to which Steiner often referred as "higher worlds." These "memory-records" reportedly contain within them the resonant memory-fields of all human experience, the truth or validity of any item within such "universal library" being a matter of opinion.

Filling in the Gaps

In 1995 I was introduced to the HeartMath Institute, a small research group (long since doubled, if not tripled in size) in the Santa Cruz mountains of California. There I found an impressive body of bona fide scientific research into the heart-brain dialogue, going on in many areas, with visiting neuroscientists and cardiologists of repute (including neurosurgeon Karl Pribram), in and out of that mountain laboratory, building an equally impressive catalog of additional discoveries and information.

By the mid-1990s, HeartMath had compiled most heart research from around the world, including the extraordinary work at the University of Arizona in Tucson. Through new scanning and imaging devices, a raft of scientific materials has given all that is needed to verify, confirm, and even clarify the ancient Shaivite scholars' explanations of the heart as center of our world. All this evidence indicates that, among a wide range of astonishing capacities, the heart does, indeed, contain a series of large and powerful neural ganglia that function as do the neural-fields of our larger "head-brain," the relationship of the two being a primal key in our life and creation.

Grand as all this was, some key ingredient was missing for me: I looked within this technical and scientific framework for a niche wherein my personal inner experiences could find room. For at critical periods of my life, events happened to me that fell—and still fall—outside all these neat systems. And I wondered where I might find a place in the scheme of things for my innermost and deepest experiences and convictions.

My encounters with Robert Sardello, Cheryl Sanders-Sardello, and their School for Contemplative and Spiritual Psychology brought together for me missing links in this strange loop phenomenon not found elsewhere. Until then I still did not know the critical role we each play in both our personal life and the life of our good Earth, whose Sophia spirit hangs in the balance with us today. Sardello's "spiritual psychology" and its heart-orientation proved to be not just another neat synthesis of previous systems, but an opening of the "future flowing into the present" I had but dimly sensed in my early years of reading the action-accounts of Jesus, stripped of the religious trappings of church and dogma.

Tracking the Movements of the Heart

Knowledge of the heart as discussed here concerns a radiating force or electromagnetic field arising from or through our heart (see figures 5.1, 5.2, 5.3, 5.4, pages 68–70). Offering new perspectives, this heart-field is often called heart "energy," as though it were of the same nature or makeup as gasoline, nuclear power, or the like, but such terminology doesn't fit. Alternately, poet Dylan Thomas, speaking of "The force that through the green fuse drives the flower, drives my green age," expresses this heart issue wondrously well, as do the terms flow and resonance.

This field, emerging out of our heart and surrounding our body, is a spiraling form called a torus (see figure 5.5, page 71). This torus the heart flow creates is a stable formation that can arise through, from, or in a wide variety of nonsolid materials, and in the heart's case, involving electromagnetic properties. The water swirling down in a whirlpool or a lavatory drain creates forms similar to a torus, and the circular rush of blood into the chambers of the heart creates a vortex similar to a torus form. So also does the pulse of blood spiraling through the major blood vessels; these vessels are reportedly "grooved" as in a rifle barrel to assist in the spin of the blood's spiraling flow.

Electrocardiograms (the familiar "EKGs" made in medical offices and hospitals) have charted the general nature of this heart-field for years. In

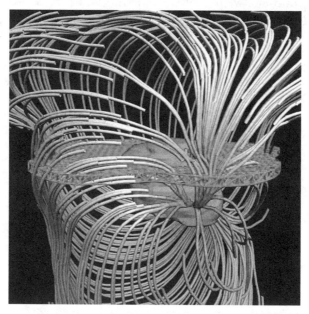

Figure 5.1. Computerized electromagnetic image revealing the forces arising from a living heart. Note there are two points from which two currents of force arise. From one point the currents curve down and up again, behind the heart—completing one in an endless array of loops pouring out from the heart. The downward currents and their corresponding upward currents move within a designated frequency range to form the first layer of a field of magnetic force, culminating in nestled torus fields (see figure 5.4). (Courtesy SCI Institute)

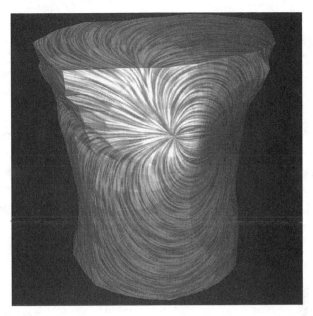

Figure 5.2. Another "live" computer image of the heart torus field, indicating patterns similar to those in figure 5.1. (Courtesy SCI Institute)

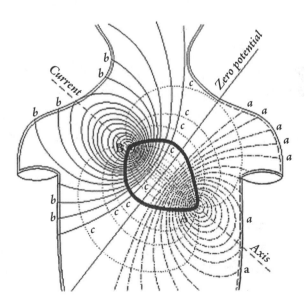

Figure 5.3. This remarkable image of heart torus fields was generated by Augustus Waller in 1887, using rudimentary electromagnetic measuring devices.

Figure 5.4. The threefold torus surrounding the human body. The innermost and possibly most powerful torus, infusing body and brain, is essentially "physical," registering physical sensations, which directly impact the body-brain. The middle torus field encompasses emotional or relational forces and influences, while the third, outermost field expands to spiritual domains.

the early 1990s, the HeartMath Institute began collecting research on the subject, as well as initiating research of their own. HeartMath's work eventually attracted neuroscientists and cardiologists from a wide area, and the field of neurocardiology—the study of the neural or "brain" components of the heart—has been growing ever since.

Such electromagnetic configuration as this torus from the heart surrounds and "embraces" our body, and electromagnetic images of this have been revealing. Of particular interest, these individual formations from our heart are essentially of the same makeup as our planetary torus-forms arising from Earth's core. These planetary flows arc out and back at the magnetic poles, creating the ionosphere embracing the body of Earth, and extending far beyond (see figure 5.6). And here, in this similitude between heart and Earth, another aspect of our heart-story appears.

As Earth's magnetic fields stream out into space, so the heart's electromagnetic field streams out beyond our body, both for

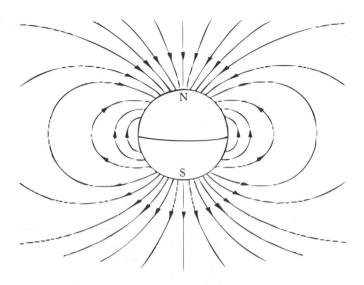

Figure 5.5. The two torus forms arising from Earth's magnetic poles, creating the ionosphere and extending indefinitely beyond.

Figure 5.6. NASA electromagnetic satellite image of Earth. Note the direct similarity of form with the torus forms of the human heart, as well as the streaming electromagnetic currents branching out from the poles, connecting Earth with similar forces in the Sun and solar system at large. The myriad semi-torus figures creating a potpourri of chaotic e-m fields immediately around the Earth, suggest chaotic patterns—in part human generated—as discussed in the text.

indeterminate distances. Further, these magnetic waves of heart and Earth fuse or merge in varying ways, depending on the "coherence" of our heart's field. And herein our story takes on deeper implications.

Coherence, in this case, means regularity or orderliness of those wave-frequencies. Given that coherent regularity, heart's field-frequencies can, and automatically will, mesh or fuse with correspondingly regular frequencies such as those surrounding the Earth—or those of other people in a similar state of coherence (see figure 5.7).

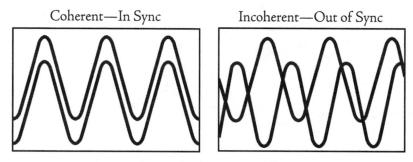

Coherent—In Sync   Incoherent—Out of Sync

Figure 5.7. Coherent wave frequencies overlap and reinforce each other, thereby gaining strength, and giving strength to other coherent wave-forms. Incoherent wave frequencies not only cannot overlap and reinforce, but clash, confuse, weaken, and even cancel each other out.

Further, as the magnetic fields forming about the Earth radiate outward in ever-expanding wave-forms, the same effect radiates out of and from the Sun, old Sol itself, which constantly produces untold myriads, torrents, and oceans of uniquely coherent torus forms covering the Sun's surface. At the same time, even more massive radiations from the Sun extend outward in great streams of plasma—electromagnetic rivers of unimaginable density, size, and power—which move into and "embrace" the entire solar system (see figure 5.8).

Coherent heart-fields within us individuals merge into those of Earth's radiating fields, which, in turn, merge with those of the Sun. Thus, we are incorporated into and reciprocally interact, directly and indirectly, with this whole solar system, to the extent our heart-fields are coherent.

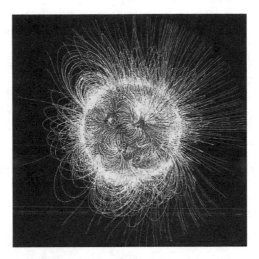

Figure 5.8. Electromagnetic image of the Sun and its incalculable outpouring of torus-like wave-forms. These streaming rivers of e-m power, known as plasma, embrace the entire solar system, meshing with corresponding coherent streams coming from the planets. There is thus an enormous solar exchange and "re-charging" within the solar system, one of the great relational power structures of creation as we currently understand it.

The Map and the Territory

A Cautionary Note

The assumption that the phenomenon viewed in this electromagnetic fashion is itself (just) electromagnetic, is, however, questionable and falls short of any final "truth." Granted that our images by which we are privy to such otherwise invisible phenomena are labeled electromagnetic, since the devices we have devised to make such imaging are electromagnetic. But to assume that the phenomenon revealed by such means is of the same form or nature as the device revealing it is simply wrong.

Were we to find evidence of some creature who comes out only at night and cannot be "captured" on ordinary film, we might use an infrared camera to photograph its activity. But we would not then assume the creature viewed to be infrared itself, that is, constituted or made of some infrared substance or material. Similarly, our electromagnetic imaging of this heart-flow reveals only a "virtual reality" of what is occurring, not the real thing as itself—the real thing is our self and our heart's nature. These are not of themselves visible. Self and the nature of heart's being are resonances found within physical expressions such as electromagnetic fields, but resonance in this case is not of itself a physical—and therefore measureable—substance; rather, it is a felt experience. To equate the two is misplaced concreteness, and leads into error and confusion. The potential for such confusion is high, and should be borne in mind whenever the term "electromagnetic" is used.

The old-line semanticists harped about the map not being the country mapped; or the word for something not being that something as-itself. In the same way, the technical phenomena giving us a "map" of that "flow from the heart" are not identical with that flow, and far from a complete explanation of it. Our logical guidelines concerning real, and facsimiles of the real, can get confused in the deluge of virtual-reality gadgetry we have created. The same error is made concerning mind and brain, which are a strange loop wherein each gives rise to and is necessary to the other—but neither *is* the other.

The Holonomic Heart in
Its Mirroring Loops

The torus organization of heart's field not only tends toward a self-generative stability; the formation is also holistic, or *holonomic*. This means any portion or aspect of the field would contain the "information" of the whole form, could such a portion be divided out, which in actuality it cannot be. A whole is just such that any portion or "part" always implicitly replicates that whole. But this holonomic torus is a movement, or flow, with no more "parts" than a breeze. And accordingly, just as a breeze can fly a kite, this flow from the heart can fly any number of miraculous creations.

A workable example-analogy of this holonomic effect is found in holographic photography. Should you make a holographic photograph of a tree using a glass plate rather than celluloid film, and then break that glass plate, each shattered part would contain the whole image of the tree found in the original glass plate, no matter how tiny the fragment. The catch is that the smaller the fragment, the "fuzzier" or less clear the image. Nevertheless, the entire original image would still be "there" in that fragment. The clarity of such fragment can be restored more or less, with the proper instruments and approach.

Such part-whole imagery is a workable analogy of what our life is all about: manifesting the clarity of the whole from our apparently insignificant and currently fuzzy fragment called "self" is the evolutionary impulse driving us from conception.

The torus from our heart, as well as those from the Earth and the Sun, is an expression of a single unified, resonant nature. The information of that totality, with which our heart's radiations are designed to fuse, is thus automatically an integral aspect of our own, functionally individual heart, as our heart is of that larger torus. Our individual expression, then, contains the same information as that of the totality. But like the holographic fragment, that totality within us takes quite a bit of time—namely a lifetime of development—to clarify and bring into sharper focus, and thus be apparent to our waking consciousness. This possibility acts as a goad to our life from our beginning, if unbeknownst.

Again, Blake is instructive: "More! More! Is the cry of a mistaken soul, less than All cannot satisfy Man." That "All" is actually what that very fragment of our original image known as "self" holds within it, and awaits and expects, even lifelong, as clarification of its own whole image.

SIX

SCIENTIFIC
PERSPECTIVES OF
MIND-HEART AND
RESONANT FIELDS

Turmoil in the Heart's Broadcast

A major source of our personal heart-field incoherence—the "fuzziness" alluded to in the holonomic perspective of chapter 5—is found in emotional turmoil, upset, or imbalance. *Emotion* is a term we use for the quality of relationships taking place, as within our body parts and/or our self within the larger body of people and planet. While often considered to be "only psychological," emotion proves to be solidly biological as well, as first spelled out in Candace Pert's classic study, *The Molecules of Emotion*.[1] Pert's Nobel Prize–winning work clearly shows emotion's molecular effect on our biological system and vice versa. In like manner, our emotions have been found to profoundly affect our heart and its torus field, with positive emotions bringing coherence, negative emotions incoherence. That such reciprocal interaction enters into further extensions of field effects, as found in planet and Sun, might seem conjecture but is the case, within an increasingly complex scalar measurement.

Should our personal heart-field be incoherent (of irregular or chaotic frequencies), then coherent merging with Earth's fields—or those of other people—cannot take place, since an incoherent wave-radiation cannot mesh with or function as an integral part of coherent waves or fields (see figure 5.7, page 72). Incoherent wave-forms, rather than fusing or merging, act in a counter fashion—clashing, even canceling each other out, rather as a short-circuit in a radio prevents that radio from picking up a station's broadcast, no matter how powerful and clear that broadcast might be.

Heart's Intuitive Foreseeing

A number of older, materialistic-scientific beliefs have been challenged by recent discoveries, stemming from more than a decade of electromagnetic recordings of Earth's electromagnetic (e-m) field output. Particularly significant is an ongoing study initially conducted by Elizabeth Rauscher and William Van Bise that monitored the Earth's magnetic fields in recordings made roughly a thousand miles apart. When compared, these two records not only matched, but showed that any physical disturbance within or on our planet, such as an earthquake or tsunami, clearly registered as characteristically marked changes in the configurations of those magnetic fields of Earth and her torus forms radiating from each magnetic pole. Certain patterns within such changes were found only in connection with some form of chaotic physical disruption on or in our planet.

We might say these recordings were electrocardiograms of Earth, whose pulsations from her magnetic poles offered a bit more for research than just a "northern-lights" sideshow. These changes in Earth's magnetic fields were later found to be strikingly similar to changes in our personal heart-fields when we are subject to negative emotions, as studied for years at HeartMath Institute. Further, these electromagnetic disturbances, clearly visible in e-m pictures of the heart, are detectable in e-m images of Earth's own torus-fields obtained by NASA from satellites. Those magnetic waves at Earth's surface, and even those arcing out short distances above the Earth, apparently shift from coherent to incoherent, according to phenomena on or within Earth herself, while Earth's great torus streamers going out into space are apparently unaffected by these more modest surface phenomena (see figure 5.1, page 68).

Research into heart-emotional interactions at HeartMath Institute discovered direct connections between our emotional responses and the frequencies of our heart and its torus field. Their research followed a direction set by Dean Braden and Karl Pribram investigating galvanic skin response (long an enigma). Following ever-more extensive research, these heart-fields within and from us were found to produce a completely unexpected intuitive, precognitive "foreseeing," detectable within that heart-torus. This foreseeing anticipated *future* laboratory-induced emotional upheavals that proved to be negative to that person, events producing incoherence in that person's heart-field.

Such precognitions were, in effect, forewarnings, since they occurred in the physical laboratory setup *in advance* of the actual event as visible to the subject or research people (see details in appendix A). Since all this research involved electronic machinery functioning independently of direct human manipulation, many an applecart of academic assumptions was upset. Such a provocative and counterintuitive phenomenon resulted in HeartMath running some 2,400 trials on this one anomaly—precognitive intuition of the heart—before publication of their first of many papers on the subject.[2]

In the same way, and equally surprising to Rauscher and Van Bise, ructions in the physical Earth are not just reflected in the Earth and her ionosphere, but also *in advance of such disturbances on an actual personal, physical-sensory level.* This was discovered when the scientists found both sensing stations showing the same shift of planetary frequency fields occurring in advance of earthquakes in general.

But far greater implications-anomalies are found in the 9-11 case, for that Twin Tower disaster, which occurred on September 11, 2001, was clearly forecast in these two scientists' recordings, showing the same significant changes in the Earth's magnetic fields and ionosphere as found in such severe quakes as the 1980 Mount St. Helens eruption or the later tsunami quakes in 2004. That the 9-11 event showed up in the magnetic images was in itself enigma enough, but, far more significantly, the 9-11 event was also *recorded in advance of the actual physical calamity itself,* as had been found with earthquakes.

As mentioned earlier, such patterns of negativity within our body, brain, and heart interactions had already been induced in subjects and recorded in HeartMath's laboratory in those many hundreds of trials. These clearly showed intuitive forewarnings by the heart of some actual material laboratory occurrence, designed to test such result in each case. This information, when published, led Rauscher and Van Bise, recognizing the significant similarities with their own work, to join forces with HeartMath. The self-evident hypothesis was drawn that incoherent or negative emotional waves—easily induced on a small scale in the laboratory—could, with sufficient magnitude (that is, from many people being so induced simultaneously), result in similar incoherence in Earth's e-m fields, as was recorded when the 9-11 disaster took place.

Media and Earth-Reactions

In previous, pre-electronic–media times, emotional reactions to local disasters took place only within a local population. One group warring on another, for instance, remained localized within that specific area simply by the comparative isolation and distance between communities. Calamities in one area obviously could not activate an emotional reaction in some far-removed locale simultaneously, and even such information arriving later and at second hand in far-flung places would be too sporadic and isolated to build into a significant "negative force."

Widespread, synchronous emotional incoherence can and does

continually take place today, however, through electronic media. This now saturates every inch of Earth's surface and atmosphere, from Arctic snow-fields to New York, London, the jungles of Africa, steppes of central Asia, or wherever. Thus 9-11, sweeping up every form of media for days on end, repeated ad nauseam and witnessed by virtually the entire population of our planet, brought about surprisingly powerful emotional reactions in those populations. Even in such far-flung climes as Australia, reports were of people rushing out into the streets to share their distress, as the images played on and on. All of this planet-wide attention—and resulting emotion—created a negative field-effect of serious, planet-wide force, *indistinguishable from other planetary physical disruptions and temporarily bringing corresponding shifts in Earth's magnetic fields.*

The issue lay not only in this corresponding planet-wide unified response from upwards of six billion humans, bringing about an emotion-laden negative field-effect, but also, as our original two scientists had found, the shift taking place in their equipment had registered on those recording devices—as had earthquakes—*in advance of the physical disaster itself.* Aye! There's the rub—which involves an even more critical enigma than the other phenomena. Resonance, as a feeling of soul or spirit, can employ or accompany the "molecules of emotion" expressed physically, much as mind can employ or accompany brain, although the two are not identical.

Collective Emotion and Chaotic Attractors

A deeper examination reveals that aspects of the 9-11 event can be shown to have also followed the same formative pattern of a long, drawn-out buildup of cumulative emotional effects involving various segments of societies and cultures. Consider the meticulous planning—weeks and months of preliminary preparations—by those dozen or so young men involved in 9-11. They actually learned to fly huge jets, intent on the same suicidal maneuver. And they were supported in spirit and intent by the organizations and rogue nations standing behind them (and source of many a conspiracy theory branching out in many directions).

Consider their lifelong indoctrination into beliefs concerning sacrifice of self for eternal rewards (which has its strong resonance with early Christian passion). These beliefs meshed, bringing together into a single focus of intent and resolve, the similar passions of large segments of Earth's people. This larger group influence of religiously-driven but otherwise unfocused resolve or intent, was brought to single focus as a field-effect of power by these self-selected "field-attractors" (to use Ilye Prigogine's term). Such scattered forces found a core resonance, attracting and building to a climax long in process, unbeknown to the various times and cultures involved.

Carl Jung spoke of a dark specter or shadow side of humankind building up in Europe in the latter half of the nineteenth century, finally manifesting as two world wars and the ultimate horrors of the Holocaust. There, a small core of passionate believers first acted to galvanize widespread political-social discontent into a chaotic focus, giving them the grounds to act as "chaotic attractors." Such attractors galvanized millions of people into a unified if disastrous response, acting to lift their unfocused chaos into ordered chaos, at the cost of well over a hundred million human lives when all was said and done.

Similar powerful forms of negative emotional energy still build in our planet today, stimulated and magnified by media. In the United States, "hate radio" involves some two-thirds of all broadcasts, feeding on and inflaming widespread public discontent.

Variations of this flood Earth's surface and atmosphere constantly with comparatively smaller and varied shifts in the coherence of Earth's e-m fields (see figure 5.6, page 71). Having set into motion a negative feedback loop, which continues in reciprocal fashion, however, with more and more people affected for ongoing lengths of time, ever-greater cumulative negative and synchronous responses spill far beyond the bounds of any country or religion.

Darwin's Formula

In the context of these ongoing global-scale negative feedback patterns, it is useful to remember Charles Darwin's claim that any action repeated long enough will tend to become a habit, and any habit repeated long enough will tend to become an instinctual, genetic reflex—accepted as the human condition. Consider, for instance, how entrepreneurs have, since the mid- to late- 1950s, employed (unwittingly or not) what is termed "startle-effects" into television programming, begun when television viewing was found to induce varying forms and degrees of catatonia, particularly in children. Young children still tend to go catatonic in front of back-lit screens in general, as do adults in varying and less extreme intensities.[3]

Media-induced startle-effects are created by a variety of arbitrary, abrupt, incoherent, and nonlogical extremities of contrasts in light, sound, and general imagery in televised content (as compared with stable natural settings). These startle-effects result in unstable, shifting visual-auditory tapestries not found heretofore in nature or ordinary daily situations.

Such a montage of visual-auditory shifts as now occupies television triggers or alerts our primary sensory-motor systems into action. These old defensive functions spring from our primary "world-brain," as Karl Pribram called it (also referred to herein as reptilian brain or hind-brain), giving rise to all human experience. These primary neural structures are unable to find a coherent, stable image in television's high-density, visual-auditory flux. So that primary brain alerts, by means of adrenaline "flight-fight" hormones—particularly cortisol— the viewer's higher, logical brain systems to "pay attention" to this primary sensory intake, which is tacitly perceived as threatening. So aroused, we viewers are loathe to take our eyes off that screen, even if we hate the programming, not aware that our alert-reflexes prompt continual if minuscule "adrenaline rushes" of this cortisol, which trigger our defense alert to lock our sensory system onto such unpredictable signals, in order to deal with any danger inherent within them.

Our old sensory-motor (reptilian) system, having no logical reasoning processes of its own, depends on the later evolutionary additions to the brain for these very much higher "reasoning" processes. Thus, in television viewing, with its ongoing startle-effect, the entire brain, "lower" and "higher," functions on behalf of that lowest, most ancient "reptilian" or basic survival-sensory process. This produces a "top-down" reversal of evolution's ordinary "bottom-up" organization of our neural apparatus, and is yet another form of devolution—compounding and amplifying the nurture-deprivation phenomenon surveyed in earlier chapters, and making us ever more reflexively subject to the negative impacts of global e-m waves as discussed in this chapter.

Sperry's Early Micro-Movies

In early micro-images of neural cells, Roger Sperry found that releasing into the brain the most minute trace of cortisol (as in television viewing) brought an immediate "explosion" of new dendrites and axons sprouting out to connect neural fields with new networks of connecting links, to cover all contingencies within its sensory-world, as with emergencies in daily life.[4]

If such startle-effects happen very often, the parasympathetic nervous system does not have time to counter this sympathetic nervous system response. Hormones released by the parasympathetic system remove cortisol and any excess neural connective links made in that instant alert. The "excess" are those links not involved or needed in the actual emergency itself. And, of course, the emergencies involved, though fictional and unreal, can produce excessive cortisol and neural connections to the point of "congestion" in neural fields.

HeartMath research shows that a single negative experience can upset the hormonal balance of sympathetic-parasympathetic systems in us for varying periods before balance is regained. Unaware of all this internal action, we adults eventually learn to compensate for and even override our continued startle impulses, *paying them no further attention*. Such hormonal actions continue, however, on primary brain-body levels beneath our awareness.

The Primary Image
Is Always True

Although this hormonal process does go on beneath the threshold of awareness, on heart and brain wave recordings such shifts and loss of our "higher" brain-mind to the apparatus of the "lower" become apparent. As physician Keith Buzzell points out (from his decades of research into a child's television response), to that old sensory-motor (reptilian) brain in our head, the image is always true.

By four or five years of age the child is aware of the discrepancies of imagery on television, and will even assure the parent, "It's okay, Mom, its only television." This is the child's "high-brain" rationalization, but the cortisol output from that primary-sensory brain continues unabated in that child, doing its damage. And this startle-effect is lifelong, as true for the eighty-year-old as for the child, suggesting hypothetically that immersion by the eighty-year-old in such imagery and the ongoing cortisol over-production may be involved in the recent increases of Alzheimer's disease, television viewing being the major preoccupation of many elderly and retired adults.

As well, this hormonal activity is only a surface effect of electronic media; far more takes place on subtle levels. As Buzzell points out, by the time our "high-brain" responds to the actual imagery of an alerting primary signal from our "old brain," millions of neural responses throughout brain and body have fired into motion, even as our higher rational processes (of which we are personally aware) accommodate to the scramble, unaware of the fuss "below," which continues unabated. And all of this, we should remember, is occurring collectively as well as individually.

More! More! (Again . . .)

Since our sensory systems eventually habituate to such false-alert prompts, an increase in intensity and frequency of such startle-prompts became more and more necessary. Throughout the early decades of television viewing, this increase continued, demanding that our attention be caught and held by the programming. Many a current young viewer today would find early 1950s TV insufferably dull. By the 1970s the ongoing alerting-action had become increasingly intense, and by the 1990s the average action on the TV screen would have struck a viewer back in the early 1950s as sheer bedlam making little sense (as such confusion does to this writer, who raised two families while never allowing such a device under his roof).

In summary, our very survival nature tends to lock our ancient sensory-motor system and its ongoing attention into such televised instability, since, being indecipherable or unpredictable (or both) to that oldest sensory-brain system, such an environment continually activates our higher neural systems (our more recent evolutionary parts of brain-mind) to attend these more primal environmental signals. Thus, our higher brain systems begin to serve the lower, almost from the beginning, in a reversal of evolutionary organization. (I have more or less repeated this outline-of-action because of our tendency to disregard it as technical minutiae having nothing to do with our real life, when it actually has a profound effect on multiple levels, both personal and global.)

Media Drug Lords

Such ongoing startle-alerts create in viewers an addiction to the high-density startle-effects employed, with their resulting adrenaline-stimulus overloads. Once conditioned to such intense stimuli and the resulting adrenaline overload, if deprived of that or a similar level of stimuli and/or overload, our sensory system undergoes a form of sensory deprivation.

Then ordinary bucolic, natural settings—such as forest or mountains—can bring boredom, restlessness, or even distress in younger generations.

These physiological responses involve the Reticular Activating System between the sensory-motor and relational systems in the limbic brain. This "RAS" closes the gates of the sensory system, allowing us to sleep, while opening and activating it to awaken us. Subjected to near-continual high-density phenomena from early on, when there is a lack of enough stimuli to maintain an established hyper-alert state (such having become our norm), a natural setting tends to bring initial boredom and eventual anxiety-distress, since our primary sensory system doesn't receive sufficiently strong stimuli to maintain full, active consciousness.

Thus, we find the restlessness of young people and their resorting to constant hyper-activity, earphones, loud music, high-tension interactions with each other, hyper-fast automobiles, and so on, and we wonder why it is getting more difficult to get even kindergartners to "pay attention," as schooling (perversely and increasingly) requires. (For which we blame the schools, teachers, methods, and so on, all of which are products of the same brain-altering process and automatically perpetuate it.)

Bending the Twig From Its Beginnings

On average, a six-month-old infant in the United States spends two hours daily in front of a back-lit radiant-light screen, as found in computers and televisions. Back-lit screens such as TV and computer screens produce radiant light, found only in sunlight or fire, stimuli which carry no environmental information. Visual environmental information, objects and actions of a world, can only be found in reflected light, and our primary visual system finds no information in radiant light other than the raw fact of sunlight or firelight.

This means that due to the thousands of hours of back-lit radiant light the infant-child experiences, he or she cannot form any visual information around which his or her world-structures can be built up (what Jean Piaget calls our "structures of knowledge"). Artificial light can reflect off objects and be cognized visually. Back-lit sources confound and confuse the budding sensory-motor receptors, whereby these infants quickly go catatonic and are reluctant to move their eyes from the screen, which makes TV the world's safest babysitter, physically, while warping brain development.

By five years of age that child will have spent some five to six thousand hours in such virtual-reality flooding his sensory system, while undergoing a seriously insufficient exposure to natural environments. Building a full sensory pattern of a real world will be compromised, and the young person will be subject to boredom and a feeling of isolation in ordinary natural settings. Some such natural environment similar enough in nature to the one in which brain formation took place in our evolutionary history must be provided as a stable nucleus for early brain development and its "structures of knowledge."

No five-o'clock alcoholic happy-hour acts in any more addictive way than this adrenal-cortisol overload and the body's coping with it. According to early studies by the medical school of the University of London, cortisol overload is a major cause of many modern diseases, particularly cancer. Such overload is an automatic adrenaline by-product of television viewing in general, with computers, cell phones, music players, and the like creating a flood of new electromagnetic stimuli, each with their varying and noncohering e-m wave-forms all adding to the general discord.

Precognitive Thresholds

Consequently, the negative wave-interferences registering in our ionosphere and reflected by us in everyday life, as found in that NASA e-m photo of Earth, is understandable (see figure 5.6, page 71). Recall the rigorous series of experiments at HeartMath Institute, originally exploring galvanic skin response and various neural signaling. This research expanded until clearly revealing that well ahead of laboratory-induced negative events, the heart-brain signals in those wired-up subjects displayed a direct and rapid intuitive foreseeing of the negative event that was going to take place.

These experiments also revealed an intriguing dilemma: the warning signals the heart sends to the brain were not *consciously* perceived by those wired-up individuals taking part in the experiments. The foreseen event had to take place on a full sensory level on the computer screen (in "real time") to affect those individuals' conscious sensory awareness, even though both heart and brain graphs clearly registered the *precognitive* sensory signals. Clearly implied is a parallel between what happens on an Earth-wide physical scale with what is also happening within our heart-brain—though we perceive it not, until after such event's physical manifestation.

Mind, the Last to Know

That Earth-wide effects can occur from the combined emotional reactions of a sufficient number of people is an enigma that typically falls outside our present-day common domain. Although such intuitive-precognitive capacities of both heart-brain and Earth can occur without our direct personal awareness of them, and are perceived by us only after the fact of their occurrence, the failing lies within mind's domain, not the sensory system as a whole.

Bear in mind that while body senses may sense as designed, mind lags behind—the "last to know"—and we can rightly ask, why?

Why is mind—our personal awareness—like the deceived spouse, the last to know concerning our very own heart-brain interactions, as in the HeartMath laboratory series? Even though in actual daily life these are signals that might be critical to survival, our mind's ignorance of what appears to be a vast intelligence and wisdom of heart-brain-body or of heart and Earth, or both, is the issue. At some point a disjunction between mind and body-brain-heart-Earth seems to take place.

Such is not the case with animals, however, who respond to heart-brain intuitive signals as automatically as any other instinct, as in the oft-reported instances of animals—both domestic and wild—clearly indicating awareness of impending earthquakes and other disasters. The odd reason for our lack of such "animal knowledge" may lie in the fact that animals seem to have no dualistic "mind," no reflective cognitive process. While they have a distinctive consciousness, heart-brain precognitive awareness and survival instincts to respond accordingly, there is apparently no intervening mind through which such signals must pass to be "checked out" on some personal, introspective, abstract-cultural or logical level, as with us humans.

So it may be that the mind, our uniquely human development, intervenes in or deflects such responses that our body-wisdom would otherwise make, leaving us open, without recourse, to whatever results occur until after the fact or initiation of any given turbulent or traumatic occurrence. Mind may always thus be a bit behind brain-heart, perhaps never catching up. This disjunct between mind and heart can, however, be bridged and overcome, an issue we will delve into as we go along here.

SEVEN

NATURE'S PLAN AND CULTURE'S CONNIVING

The Breakdown of Nature's Plan: Conflict of Old and New Evolutionary Structures of Brain

This book defines "culture" as a negative force that confines, constricts, and limits humanity. Nature's positive and intelligent *intent* as given through evolution has been overwhelmed and short-circuited by a negative culture's *intentions*.

Nature's intent for culture is to give the grounds for a creative ability arising from the mind's capacity to imagine—to create internal images not present to the sensory system, and then actualize such images, giving them a presence in our sensory world. A natural flowering of our social instinct and the bond that nurtures and fosters creativity could bring a new order of social reality transcending the limitations and constraints of this present one, wherein the heart would find its next level of evolution.

As an analogy to the negative culture we find ourselves in, consider the culturing of an organism in the laboratory, as when a medical researcher comes across a microbe he wishes to study. Determined to separate his specimen from any unwanted influences, he makes a culture for and/or of the specimen. Putting it into a beaker or test tube, he adds a solution suitable to feed and keep it alive, controls the temperature of this arbitrary and restricted world, and observes and draws conclusions from what he terms the specimen's actual, that is, natural, behaviors.

In a true sleepwalking manner, we humans are specimens caught up in an elaborate cultural test tube or beaker of our own making. As Gregory Bateson pointed out in his book, *Mind and Nature: A Necessary Unity,* we are Nature, and what we are doing is what Nature is doing (if at second hand, once removed).

Thus, any and every effort we make toward a "new order" of our condition is but *a different mixture of the same old materials.* We watch wearily as, one after another, our brilliant inventions, hailed with such excitement and hope for the better world they promise to bring, end as fatal traps wherein we are again and again hoisted by our own petard. So culture, with its mad-scientist servant-master, tightens its stranglehold on the spirit as well as the body, leaving a reign of madness and a dying world in its wake.

Of course, the word "culture" has its equally powerful, meaningful, and positive definition, summarized (above) as a shared social coordination that stimulates and fosters development of creative pursuits over and above ordinary, instinctive survival skills. In this positive aspect of culture arise our great arts that can transcend all present levels of knowing: the musical works of Bach, Mozart, Beethoven; the paintings of Michelangelo, Raphael, Leonardo; the altruistic efforts of St. Francis and of the Knights Templars and Cathars, who gave us so great a work as Chartres Cathedral; the great poetry of Nemerov, Whitman, Blake, Milton, Shakespeare—any of which are, in this author's opinion, far greater achievements of humanity than walking on the moon or building atomic bombs.

We might define a creative culture as a balance of Simone Weil's "grace and gravity," wherein grace arises from creative expressions of the human spirit, gravity from the necessities of physical existence. These two polar effects are graphically played out in our own head, as displayed in the conflict of our "reptilian" hind-brain functioning automatically for survival and the new prefrontal cortex, avenue and translator for creativity and freedom, struggling for expression and ascendancy, with our sense of self at stake.

The Neural Tube and Our Sense of Self

Our sense of *self* is linked, even entangled with, *mind,* but they are not identical in actual usage. Self is also entangled with the word-concept *soul,* a far more controversial issue than mind; all three—self, mind, and soul—are entangled, arousing different connotations and subtleties. (*Infusionism,* for instance, was a theological term for the theory that the human soul is "infused" into us at our conception or at birth.) *Self* has vast connotations in our usage and references to it, but is nowhere near as elusive and loaded a term as *soul,* a slippery issue I leave hanging here on behalf of the word *self,* which has an immediate, intimate, and near-ultimate meaning to us; as pointed out earlier, for most of us it is that which we mean by *me.*

The following proposes that self, apparently indigenous to humans, originates in the *neural tube* of earliest embryonic development, a hypothesis that takes on meaning and strength as we follow through. Just as growing a liver in our gut is inherent within our genetic system, and all "liver components" congregate in the same locale in embryonic growth, so does a sense of self grow in its respective initial locale and "matrix," the neural tube. This hypothesis is borne out in physical development, as that very early forming organ, the neural tube (inherited from an early vertebrate embryo), gives rise to a fundamental trinity of heart, brain, and spinal cord, in quite distinctive ways.

This neural tube morphs, through stages, into our functioning heart beating away in our chest (see figure 7.1). At the same time, this same neural tube is equally the matrix for birth of those neurons from which the brain in our head is fashioned (thus the term *neural* tube). This neuron-birth out of the neural tube continues along with the formation of the heart. These neurons formed in the neural tube must migrate out of that neural tube matrix and its incipient heart, into that locus for an emerging brain, in our equally incipient cranium. Heart and brain thus have a common original matrix, the neural tube, which is, I propose, to be the matrix of self as well.

This sense of self, endemic to humans, is expressed through heart and brain equally, even as heart and brain develop in their separate locales later on, in order to do their "separate" jobs. These two locales of self—heart and brain—serve both functions of self, without self losing its singularity as simply self. Eventually self expresses its universal aspect in our heart, and its individual aspect in our head, these two functioning as a mirroring strange loop. Brain and heart, having given rise each to the other, cannot be without one another. Yet neither is identical with the other, nor could they be, if the twofold system is to work. Strange loops are reciprocal processes within creation, and within that created, not one-way streets or singular, isolated functions. No man is an island, nor is a single cell, or a function, or anything else.

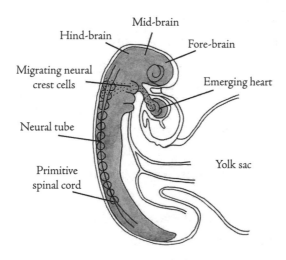

Figure 7.1. Neural tube. (Line drawing by Eva Casey)

That the neural tube is the matrix of both brain and heart, both of which contain or express and are equally centers of self, is the issue. We can speak of a self in the heart and a self in the brain as a single self, since heart and brain are themselves "twin expressions" of our singular genetic neural tube system, playing their respective roles of specialization. Self will be, or is designed to be, a heart-brain affair. We have seen how this perfect symbiosis of self-heart-brain gets derailed into two essentially separate and, on the surface of things, even antagonistic systems. This schism leads to a long parade of disasters plaguing humans, which results from just such an induced, unnatural, and arbitrary "civil war" between heart and brain. This inner-body conflict may be the final state of a divided self—before it self-destructs.

The Splitting of Self

As originally designed, the heart-brain, in its twofold locus and structure, is the means by which our concept of self can be expressed as both universal and individual, each giving rise to the other. The unfolding of their strange loop is the story of human development—or failure. *For here is the site of an eventual breakdown of this dual interaction,* a split of that singular universal-individual self that brings about the proverbial fall of man, leaving us a species in constant free-fall, in fact, with "no place to lay our head" (such as even animals have).

That heart and brain are a single reciprocal unit "separated" according to usage has a direct correlation of right-hand and left-hand specialties. The ancient Rig-Veda refers to the "two hands of God," each with its "specialty," yet functioning only in sync with each other. Some Christian texts refer to Jesus "sitting at the right hand of his God" (which of course places his God in that left-hand bracket, which fits the picture before Paul and his Greek mythical-glosses over the original Jesus took place). There are other ancient references pointing toward this singular creative process that requires a dual means of expression, all reflecting the fact that there are no "one-way streets" in creation, no isolated, self-sufficient phenomena having no connection with any-thing—not even minds, souls, psyches, or what-have-you. All phenom-ena are reciprocal on some level, since only movement exists.[1]

Earth's Pathology:
Reciprocal Relations Gone Awry

Cellular pathologists Scott Williamson and Innes Pearse spent some three decades at the Pioneer Health Clinic in London, observing six families followed through three generations. Looking for what con-stitutes the general nature of health or wholeness, this husband-wife team, after years of insights and discoveries, were led to the unex-pected conclusion that the Earth is a living sphere, a single living body (or, as poet Blake might say, the living body of a single great human).[2] The surface life-structure or "skin" of this single Earth-being consists of plants and animals as intimate, critical parts of the total planetary life—rather as our body and skin are an interdependent unit. And we, as the most advanced of all species, most profoundly affect the living body of Earth, a body we are born to serve and be served by. (Williamson and Pearse's proposal of Earth as a living being predated the similar Gaia theory.)

As Williamson and Pearse also explained, cellular health depends on a cell's integration into and cooperation with the overall coherent functioning of the body of which that cell is part. Cells continually wear out and are replaced, while some simply lose their frequency-coherence, or ability to function as part of the coherent whole. Those incoherent cells continue to absorb nutrients, while contributing noth-ing to the whole and taking no part in body maintenance (again, since incoherent systems cannot mesh with or "communicate" with coherent ones, as explained in chapter 5).

Meanwhile, these incoherent renegades divide and multiply, as all cells are genetically programmed to do, thus increasing both inco-herent cells and the uptake of nutrients, while adding nothing to the whole except more incoherence. This breakdown happens continually among any creature's cells, and is ordinarily handled with skill by the body's immune structure, dispatching these breakdowns as they occur.

Missed by the immune system—not excised and got rid of—such an incoherent cell functions *only for its own survival or maintenance,* since that singular survival reflex or instinct is all that is then available to that cell, having lost its reciprocal frequency relating it to the intelligence of the whole. This intelligence of the whole, which is its well-being, functions through (or as) the heart, but the incoherent cell—analogous to our incoherent human being—has lost its resonance with that intelligence. So that single incoherent cell survives at the expense of the cellular web to which it belongs (or did belong). The maverick cells can seriously disrupt this web, absorbing energy while contributing nothing other than reproducing more and more of their incoherent kind.

Williamson and Pearse rightly called this type of disruptive cell, acting only for its own well-being at the expense of those around it, a cancer, and concluded that the human species has itself become a cancerous growth on the body of the Earth, with each "human cell" of that "larger body" of man-Earth functioning only for its own survival. (Any similarity to unbridled capitalism as practiced this past century and elevated to near sanctified status in our pathological society, is well intended.) And, as these two cellular pathologists concluded half a century ago, we can expect this incoherent derailing to spread exponentially until destroying the whole body of man-Earth, as a cancer gone wild tends to do. (I mourn both for the body of Man, my body, and for Earth, our body—and their reciprocal life-processes.)

Sophia's Service or Its Opposite?

Robert Sardello speaks of Sophia, the Spirit of our living Earth, to be served by us as good stewards; Sophia, in turn, as our matrix, nurtures us. This is a cosmology and awareness shared by the Kogi, a remarkable, almost totally unknown civilization occupying a huge mountain on the coast of Colombia, South America; they have deliberately isolated their culture for four centuries or so, since the Spanish invasion. The Kogi voluntarily revealed themselves recently to explain to us our destruction not only of their civilization, but of the living Earth herself, warning us, of course, to cease and desist from our destructive policies. The Kogi described how we moderns no longer serve the Earth and her Spirit, but rather, by destroying her will die with her. And the Kogi, several thousand of them, are indeed being eliminated from our planet today, by our contemporary mind-set and technologies, even faster than we are eliminating ourselves.

Instructive at this point—and consonant with the Kogi's warnings and Sardello's Sophianic view—is our earlier concern with strange loops and the resonant merging of fields within fields: fields of people with Earth and the field of Earth with Sun. These show a reciprocation between the forces that power both planet and, indirectly, each of us. But our particular "part" of the holonomic totality, rather than becoming more focused, resolved, and coherent, is growing ever-more fuzzy and chaotic. Incoherent action on the planet, brought about by its corresponding reciprocal action, is reflected in a variety of ways—continually more unpredictable earth-and-sea changes and their parallel in the rise of incoherent personal and communal patterns of behavior. Behavioral changes are displayed in the mounting irrational anti-survival and destructive actions against each other as well as our Earth, gaining ever-greater momentum. (One recalls the Orwellian prediction of a war of all against all, as we read of American families arming to the teeth to protect themselves against their armed-to-the-teeth neighbors.) Such negative behavior enters into the mounting incoherence that has seized the reciprocal balances of our life and planet and feed back in ever more powerful reciprocal reactions, affecting us creatures on the surface who then affect the coherence of Earth itself, round and round.

Evolution's Response

At the core of this dilemma lies that tired and ancient issue, our fear of death, the adverse side of life, which may underlie our equal fear of both life and death, as so nakedly shown in Sigmund Freud's writings. Lower species express many gradations of survival-fear, but, having no "mind" like ours (we assume), probably do not face the daunting and haunting problem of *eventual* death as we do, a fear arising from late childhood on.

Subsequent pages here will explore an evolutionary "work in progress" which has been in process as long as we have been around. This work in progress is evolution's response to this most overwhelming of all limitations and constraints, which is not so much personal death as our crippling fear of it. Death as itself has no evolutionary challenge, but a life lived in fear of death sets up a constant incoherency that limits us severely. In this light, we can now revisit our earlier perspective: *evolution is the transcendent movement to go beyond limitation and constraint.* Now, we are considering limitations and constraints that affect the whole planet.

Creation and Evolution:
A Mirror-to-Mirror Strange Loop

To understand our own life-creation is to understand all creation, and to understand our evolutionary makeup is to understand the evolutionary makeup of the cosmos. Evolution and cosmology embrace both the broad universal and the purely personal, showing them to arise from and as the same function. The cosmos and I are of the same creative process: a mirroring loop of potential in the process of realizing itself.

Creation takes place through an evolutionary process, just as evolution is that creative process. They, too, are a strange loop. The endless expressions of creation, spreading out from us universally, are essentially of the same order as our self and its searching within. We are the only way creation can be, but we are not identically that creation. As Meister Eckhart expressed it some six centuries ago, "Without me God is not," recognizing that without God, Eckhart was not, while in no way assuming that he, Eckhart, was God—or vice versa. Eckhart and God were a strange loop, each giving rise to the other, as is the case for each of us.

There can be no creation without a creator, nor creator without creation, for the two phenomena, like self and universe, give rise to each other—neither possible without the other, and so essentially a single phenomenon with two aspects or faces—rather like our initial heart-brain as a singular self.

Stepping Out into Nothing and the Strange Loop

Creation is an endless process stochastically exploring every possibility of being (*stochasm* meaning purposeful randomness). Evolution is an ever-present urge within all created event-phenomena to move beyond the limitations or constraints of any created event-phenomena. So evolution is the transcendent aspect of creation, creation the response to evolution. And every phenomenon, every event, has its eventual limitation and constraint. There could be no creation that is final, since even the concept of finality would indicate limitation against which evolution, as is its nature, would perforce move creation to rise above and go beyond.

To move beyond limitation and constraint is a twofold process: first, to generate such movement itself, and second, to create that which lies beyond. And that which lies beyond the limiting constraints of something created comes about and is realized—made real and actual—only by the movement of transcendence or "going-beyond" itself. "Where" transcendence might go in "moving beyond" is determined by the going itself. ("We walk by falling forward, and go where we have to go . . ." as poet Theodore Roethke spoke of it.) Our "going" enters into the nature of that which we enter into and brings about *by* our going—*the very definition of a strange loop*. And herein lies the central thesis of this little argument of mine, as it did in my first work over half a century ago.

By its nature, evolution reveals all "points of constraint-limitation" in creation, and creation takes place stochastically, in a process of random profusion and a purposeful selectivity from that profusion, according to what works. Like water, evolution seeks out, through means, which come about *through this seeking-out,* a level that lies beyond its present state—the "present state" being all of creation at the moment of the stochastic movement to go beyond.

Thus, the cosmos expands-evolves, moves beyond itself, as does the evolving person; wherein lies the key to our evolution's move to overcome the apparent final limitation and constraint of death, or the fear of it—or both. Cosmos and person are of the same order, the same essential creative function, regardless of "scalar" difference (light-years or micrometers). And we *are* that creative function on its micrometer scale, or were, until we invented electron microscopes that call for a yet finer gradation or scalar measurement. No crowding down there (or here)—plenty of room for yet further stochasm and evolution.

EIGHT

DARWIN'S EVOLUTION

Since Charles Darwin's time, we rightly see life as evolutionary, and evolution as a fundamental creative function of life. As expressed in his first book, *The Origin of Species,* he explored how a single-celled creature in some primal ocean could end as the vast diversity of plant and animal life presently on Earth. He observed how myriad life-forms went beyond their original state through mutation of their genetic material, selectively filling in the new form from the possible profusion of matter found.

In the years since *The Origin of Species,* we have discovered through the electron microscope the double-helix molecule we call DNA, with its corollary RNA, and the wonders therein. This can be seen as a key element in Darwin's proposal, and a monkey wrench in certain of its ramifications.

DNA: Thing, or Activity?

We might consider DNA (mentioned in this book's introduction) as a phenomenon bridging the gap between reality as matter-substance and reality as a phenomenon of mind—an experience or venture as in imagination or mathematics (neither having substance but both having great creative power).

Of itself, this double-helix making up DNA is as "next-to-nothing" as they come. It measures less than ten atoms across, far more narrow than a photon of light. This is why light microscopes never discovered the DNA molecule. It is so narrow it simply passes through a photon without registering an effect. So this elusive DNA substance-thing had to await the electron microscope to make its entry as a bona fide, electronically real scientific object.

The actual strands Nature has paired together to make this helix are made of four basic compounds repeated over and over in varying combinations. The resulting strand is some six feet long, and the number of combinations of these four compounds is astronomical. This hairline six-foot phenomenon folds up in a nearly infinite number of ways, each way capable of bridging the gap between being-as-potential and being-as-real in our ordinary sense.

A living cell is a nest for such DNA folding, each folding theoretically able to unfold and be put to work to create a different material substance or thing. Yet, were this next-to-nothing double-helix molecule extracted from each of our body's approximately seventy-eight trillion cells and hooked together end to end, that double strand would stretch from Earth to the moon several billion times. This makes for a considerable stretch of our imagination—but recall that imagination is the capacity to create an internal image not present to the outer senses, which can then be used to enliven some particular potential waiting around in limbo for expression. (Thus, poet Blake considered imagination Divine—the way we are made in the image of God—remembering it is our image that goes into the making, as with any strange loop.)

In reifying DNA as we often do, we too easily grant it a material thing-substance status we can mentally handle, like building blocks in the sandpile. That is, we too easily cross the line between mind and its imagination (our ability to create images not present to the senses, as found in mathematics, for instance), on the one hand, and catalog events imaged as actually "out there" in a supposedly real, material world of realized potential, on the other.

DNA defies explanation or definition within a Darwinian evolutionary creation. From where could such a riddle as DNA have come, lest we fall into "creation ex nihilo" (out of nothing), which is a no-no in all standard requirements for really-real as held by science-says.

Near hilarious was the Watson-Crick "trans-spermal" proposal-explanation: that the first DNA our good Earth experienced, triggering evolution, arrived on our planet by an asteroid or meteorite from some solar body that had indeed synthesized or created such a miraculous combination. According to the Watson-Crick theory, this material phenomenon-sperm, caught up in a material object, ejected out as a meteorite or some similar conveyance, traveling through space, to land on and grace our planet—as though our planet could not of itself bring about life. This embarrassing "infinite regress," similar to flying saucers or Stanley Kubrick's movie *2001: A Space Odyssey,* is entertaining but somewhat lame, coming from Nobel laureates, halos and all.

DNA: Form and Content

DNA's molecular form-structure functions like a blueprint for a building, simply a proposal or imaginative suggestion, in effect, providing the outline-directions for constructing a "thing," but no *content* to realize, or make real, that imaginative possibility. The content for the form-proposal comes from the environment. (This is admittedly a variation on the infinite regress trap, or an evasion of it.)

Mutation is a rearrangement (even scrambling) of a particular DNA form-guide, the possibilities of its scrambling being virtually infinite, as just mentioned. The content for "filling in" the new form comes from the vast profusion of life-forms, or perhaps raw, formless matter in the environment, if such befits that particular new DNA arrangement.

Only some of those mutations continually taking place survive as new life, however, as only some find an ecological niche, giving a content suitable to their form. Each mutating form can then move beyond the limitations that lay within it before undergoing such change, which is an expression of evolution, and we are each the current end-result of such form-content adaptation and change.

Stochasm's Lots and Lots

Stochasm, as we have seen, is a Greek term *for randomness with purpose.* Academic science has long accepted randomness and denied purpose, a strange mental block. Creation moves stochastically, that is, through a purposeful randomness in a profusion of potentials, vast quantities from which any particular blueprint might find its needed content.

Each of the great oaks out in my yard drops untold thousands of acorns each year, while comparatively few little oak sprouts appear the next year—far fewer sprouts than acorns. And of all those little oaklings popping up, even fewer survive. Perhaps only one mighty oak will actually make it to maturity every fifty years or so. Otherwise, the woods outside my door, hardly orderly at best, might be a real jungle.

With insects the numbers are astronomical. An astute biologist calculated that if a single fly were to survive, grow to lay its own quota of eggs, which in turn survived, on and on, within a single year the Earth would be covered by a layer of flies some three feet thick. Fortunately the ecological niche for flies has its constraints, and only the fittest of that multitude survive, maintaining the vigor and productivity of the fly-herd, irritating me but making room for all of us.

Stochasm and Embryogenesis

At birth, the human female has some five to seven million prospective eggs on her ovaries, this "egg" being not much more than a bundle of DNA. The six-foot human-DNA-molecule "folds" its infinitely flexible double-helix into that microscopically tiny molecular form, according to environmental signals of which configurations of folding might best succeed in that environment. Those environmental signals change continually over time, and between birth and maturation, that number of prospective-lives-in-egg-form (hugely profuse to cover all conceivable bases) is reduced to a "mere" three or four hundred thousand. As different environmental signals flood the mother-body, those DNA configurations not matching these endless environmental shifts are processed out: simple, pragmatic practicality for humans or flies.

When ovulation begins some twelve to fourteen years after birth, those remaining several hundred thousand bundles of DNA continue to be selected as best suited, or "matching" the ever-changing environmental conditions then present. From this pool of potential there are some eggs further "selected out" regularly, as maids-in-waiting, to be encased within that membrane we think of as the "shell of the egg" (not quite like that breakfast egg this morning), and readied for the drop down that famous tube to await her suitor. This process of "membrane" formation—completing the egg—takes considerable time, from selection from the mass to readiness for its nuptials. A lineup of a scant few dozen much-selected egg prospects are in the assembly line at any one time, undergoing their final maturation, nuptial finery coming up, misfits still being weeded out right down to the wire. In the course of a woman's long life a scant few hundred eggs at best might make it down that tube, where an even more minuscule number get fertilized, and far fewer still go the full nine months to completion of a whole new human.

As for that male suitor involved, the situation is markedly different. We human males are born without a sperm to our name and don't start production of such creatures until just about that time when human females begin to ovulate, somewhere in the mid-teens.[1] Once that male machinery is turned on, however, the numbers involved are staggering. Not only does the male produce sperm constantly round the clock (continually updating that output as current environmental signals indicate, and with quite a different selective process than with female's future-oriented nest egg); a single lucky offering the male might get to make will loose into that sacred river between four to five *hundred million* contestants for the prize. From the moment that huge swarm hits the stream, however, massive selectivity sets in, and by the time the Grail is actually reached there will be but a handful of hearty swimmers left. (That bundle of DNA called a "sperm" is propelled up that stream by a ridiculously long tail, which has its own remarkable history, as described in this book's introduction.)

But, according to new reports, which may be based on facts giving grounds for romantic hypotheses catching our fancy (at least mine), Nature has one more selection to make from even these stalwart survivors. Those few sperm reportedly form a circle and navigate around the egg (that egg being many hundreds of times larger than those tiny sperm), at which point the egg apparently decides for herself which suitor is resonant with her—that is, which best matches her own DNA patterns and expectancies (assuming such natural process is allowed to take place and no virtual reality such as a test tube has intervened). At the precise moment of proximity the egg opens a tiny portal in that membrane wall of hers and admits the winner of the prize, closing quickly lest a lesser rival should crash the nuptial party. Such selectivity does not lend itself to the usual mechanical notions we assume but indicates a "resonance" between egg and sperm that may lie outside our usual molecular-materialist notions.

At this point of entry, the long tail of that mighty sperm drops off, since no longer needed (implying it would clutter up that sacred space were it carried inside, while actually the "tail" doesn't exist as a substance-thing, only as a force or process, also detailed somewhat in our introduction). At any rate, consider the enormous selectivity involved in bringing about the conception of just one of us humans.

And the selectivity does not end at conception. Estimates are that as many as 90 percent of all fertilized eggs may spontaneously abort by or around the tenth day after conception. It may be that the neural tube begins its first preliminary cellular division-growth at that point, but whatever the specifics, if the supportive environment is not sufficient or cell division is faulty, the attempted conception is abandoned and Nature opts to try again. In such cases, most women are not aware of having conceived and aborted, since these earliest embryos are too tiny to be noticed. A woman might be "late in her period" or feel she has skipped one, never aware of Nature's selectivity at work. Other natural abortion periods seem to center around the history of this neural tube and the embryonic phases opening around the third, fifth, and seventh months after conception. If these early formations are not up to standard, Nature blithely gets rid of the attempt and clears the deck to try again. There are, after all, hundreds of thousands of eggs left and seldom a dearth of eager male donors.

Mead and the Samoans

Another example of Nature's profusion-selectivity in human reproduction is best told by Margaret Mead's studies, found in her treatise, *Adolescent Sterility*. In her early travels among the Samoans of the South Pacific, she discovered that teenagers reaching (or approaching) puberty seek out and are given total sexual freedom. Indeed, anything less had never occurred to these people. At fourteen or fifteen years of age, when that tantalizing urge begins in earnest, children begin an intense and exciting exploration involving any number of partners—an exhilarating method of selectivity for that best fitting the overall ecological niche. (The Trobriand Islanders had a "children's house" where the young could pursue their courtships, couplings, and adventures without interference.)

For three or four years this exciting and exhilarating pairing-off and trying-out took place. (Recall the intensity of your own adolescent "crushes," which flourished in spite of the criticisms and fear-laden restrictions from the negative cultural environment to which most of us were subject.) Around eighteen years of age, permanent couples began to form, which eventually separated from the group, underwent the customary nuptials and went their private ways. Why so long a selectivity period? Because Nature, being an expression of love, loves herself—which is life loving itself and loving reproducing itself. Why shut down such a lovely, exhilarating, heart-poundingly exquisite experience as new young-love discovers? Why not let it run for several seasons? Nature's loving nature constantly perfected sexuality from its most ancient lowly origins to its current perfection in her greatest creature, Humankind. So, as poet Blake asked: "How do you know but every bird that cuts the airy way, is an immense world of delight—closed to your senses five?" Nature loves sex since she loves herself, providing for it as appropriate and opportune, since it is also her great ally, keeping the wheels spinning (see appendix B for a brief report on two astonishing works by French physician Michel Odent—*The Scientification of Love* and *The Functions of the Orgasm*—which explain the universality of creativity expressed in sexuality and rightly place sexuality itself as the prime universal power and meaning).

Selection for Stability

Mead was puzzled by one major issue. In all that sexual exploration and partner-switching, no conceptions took place. No pregnancies occurred! What took place was, however, highly appropriate to the well-being of family, child, and society and, I would claim, was a most intelligent response of the heart and its guiding intelligence in Nature's plan.

Once the couple had gone through their nuptials and lived together for a while, the young woman generally decided when she should have a child, prayed to her private goddess at one of the island shrines, asking permission, and soon became pregnant.

A couple of points should be made here: these people were famous for their marvelous (if quite large) physiques and healthy lives. Their family groups were stable. They had sown their wild oats thoroughly in those teen years, and had no need or interest in trying to be teenagers forever. Nature, having achieved her primary selective needs for the appropriate gene-matching and generational refilling of the ranks, could enjoy her great orgasmic sport at leisure. Further, every child conceived by such gene-matching, making for the best genetic results, was born into a compatible, settled, harmonious and stable setting where every child was wanted and totally nurtured, breast-fed for three to four years, and largely carried by the mother for that first critical year or so. Further, stair-step children (one born every year or so) did not occur. Offspring were reasonably spaced, a sibling coming along only every four years or so, were the first child a male, three years were it a girl.

Nature's plan in these matters seems to be that so long as a mother is lactating she is essentially infertile, and these natural mothers nursed their boys longer than girls since males are far more fragile than females, require a longer and more protective nurturing, and do not fare well if a rival comes along too soon and competes for that breast.[2] (Males also account for the majority of aborted, stillborn, or defective children today.) At any rate, these Polynesian people were in tune with and expressed Nature's intelligence (intelligence being the capacity to move for well-being, which comes spontaneously without thought or effort, from the heart). And, as went the popular song of the '30s (disapproved by my Baptist forebears), by simply "Doing What Comes Naturally."

While these balancing factors were found in many preliterate "natural" societies, they would be a disaster were they attempted in a religious technological culture such as ours. You have to buy the whole package in such balances of Nature, not pick at her piecemeal while at the same time violating her at every turn, as is our practice today.

More, More for Less

Overall then, Nature functions through stochastic profusion, overproduction, and rigorous selectivity (as found with egg and sperm), according to need and in a rich complex of intelligent responses. We look out on a night sky with millions of suns blazing, among which untold number some may have the right conditions to grow themselves some planets, and of these millions of planets some will have just the right conditions for growing some life. In this infinitely vast universe, virtually an infinite number of planets and lives continually emerge and go their way. Darwinian selectivity and survival of the fittest have to do with the complexities of a most complex and yet simple-intelligent design of staggering brilliance, bringing about the most successful life-forms possible.

———

NINE

DARWIN II

Death and the Evolution of the Mind

Following Darwin's second and greatest work, *The Descent of Man* (850 pages of fine print, on which Darwin spent the latter part of his life, and which has been radically ignored), we find that evolution indeed moves to go beyond even the "ultimate" limitation and unyielding barrier—death, and our fear of it. We can further plot out the nature of that barrier of fear and the kind of mutation-selectivity involved in moving beyond it—an extension of the evolutionary process which took a bit of groundwork.

"Darwin II" (*The Descent of Man*) explains how those forces of mutation, selectivity, and survival of the fittest, which can account for the myriad variations of species on our Earth, cannot alone account for the appearance of the human species. According to Darwin's second offering, we were brought about by the "higher agencies of love and altruism," rather than simply being the fittest and surviving our predecessors.

David Loye brought this ignored aspect of Darwin to our attention in his excellent little book, *Darwin's Lost Theory of Love*. Consider as well such random items as Darwin's essays on "Selectivity in Relation to Sex," "Expressions of the Emotions in Man and Animals," and other writings and essays that reflect his involvement with the thoughts of

Ernest Haeckel, Lamarck, Goethe, and other biologists and minds of his time.

Consider these most telling items to which Darwin refers in his last work: "aiding the weak to survive . . . the instinct of sympathy . . . the noblest part of our Nature. . . ." Where in these words can we find justification for the tooth-and-claw, Neo-Darwinian jungle mentality that swept the academic-scientific scene in the late nineteenth century and is still rampant today, if in ever-new cloaks or disguises? These guises justify the ever-mounting travesties and demonic actions we witness daily, sanctified, in effect, through the growth and power of materialistic-technological science, capitalism, colonialism, and the ongoing domination and destruction of culture after culture. Here in this Neo-Darwinism is the very dark mind-set that paves the way for the mounting wars, death, and destruction dominating the twentieth century and poisoning the Earth today.

Humankind, through millennia of the Darwinian function of "higher agencies," with its "noblest instinct of sympathy," developed an awareness beyond that of any animal, a self-awareness leading to an intelligence lying at a light-year disconnect from anything that had come before. In turn, this self-aware intelligence led to a new function that is not physical, and thus not directly accountable to physical constraints such as death.

In this way, evolution, as revealed by Darwin, has struggled to bring about our ever-evolving human mind. Emerging out of evolution's matrix of body-heart-brain, we are continually being led to a higher evolutionary capacity, one that is able to go beyond ever-greater limitations and constraints, opening to ever-new matrices or sources and capacities. Mind (and here "mind" and "soul" are a tangled knot) is designed by evolution to move beyond ever-greater stages, just as the infant rises to toddler, to child, to adolescent, and so on, simply by living life to its fullest. No wonder the cultural forces of darkness recognized by Rudolf Steiner now mount in even greater opposition to our true nature bestowed millennia ago.

Mind Emergent

Mind as an emergent process has risen along with a field-effect, which mind automatically attracts to itself by its new capacities and passions, and to which mind is attracted by such passionate pursuits and quests. Such related fields will be drawn through "resonant" attraction—like attracting like. Mind thus brings about or gathers to itself fields of possibility that serve as matrix for that ever-expanding mind to draw on in reciprocal fashion, as it does with body-brain and heart. These new, nonphysical matrices are fields of potential that lie—or can lead—beyond strictly physical process. In like order, physical life both belongs to and brings about a continual supply of new materials, new fields, and new transcendent minds, in strange-loop, mirroring fashion.

Bohm, Sheldrake, and the Field Phenomenon

To further explore the nature of fields and how they participate in the evolution of mind, consider the "field perspectives" of physicist David Bohm and biologist Rupert Sheldrake, whose conversations in the 1980s were often facilitated by the philosopher Renee Weber.[1]

Bohm brought to these conversations his concept of *active information.* This view holds that information passes back and forth between an electron and its environment-field, in a mind-like interplay that impacts both the activity of the electron and the form of the field it operates within—an ongoing dance that clearly echoes the strange-loop phenomenon. Furthermore, Bohm's view assumed that in some fundamental sense the electron only appeared to be separate from its field; in actuality, the electron was a pulse arising from the field itself. Bohm proposed that this field-electron dynamic was applicable at many levels of reality, including the human being and the fields of influence the human being generates. Similarly, Fritjof Capra noted that in quantum physics we never end with *things,* or the safe round solid objects sought by the materialists, but only connections, influences, forces. Or, as I have insisted in my previous books, all there is, is relationship.

Sheldrake's scheme consists of the twin concepts of *morphogenetic fields* and *morphic resonance*. A morphogenetic field is an information structure that lies outside of space and time but acts within these to give form and structure to three-dimensional phenomena. The embryogenesis of a horse, for example, is understood to be informed by a morphogenetic field specific to that type of horse. In some sense all horses of that type—past, present, and future—are connected by this field. Morphic resonance is the means by which a particular horse and its corresponding field "communicate," and is responsible for the ongoing continuity of that type of horse, as well as novel alterations and further creative development.

For our current purposes, note that both the concept of a nonlocal field of information-meaning and the notion of formal resonance (in which similarity of form or vibration enables communication between apparently disparate elements) contribute to our use of the term *mind-field*.

Mind-Fields

Resonating and Mirroring

Mind-fields, as we shall use the term, are aggregations taking place through resonance between events—events of a like enough order tend to aggregate as a "field" of that order. Resonance is a form of relationship, in some cases indicating some shared origin or destination of two events of some like order. Neuroscientist Paul MacLean points out that the brain's hundred billion cells function by resonance, not by the matching and sorting of facts like a super encyclopedia. "Mental" fields, being non-local and non–temporal-spatial potentials, lie outside any substantive, concrete referent (and for that reason have long been dismissed by materialists). While "field" in this sense is a handy term for various actions or states of mind, such a field has no localization. Field-effect is, like gravity, a verb, not a noun; a process or procedure, not a product; an aggregate of potential as in a storage battery; a hypothetical grouping of related actions and/or possibilities for action.

The focus of our concern lies with this nonphysical field-effect, which is both necessary to and brought about by our emerging-evolving mind. Common logic would suggest that the evolution of mind (evolution being, again, the urge to move beyond limitation and constraint) is brought about and then spurred on by the mirroring of that mind's own evolutionary growth with its increasing awareness of its mortality. Mortality is always looming as a kind of super-constraint and limitation—bringing an ongoing "spiraling gyre," as creative-evolution proves to be.

(Some traditions consider such a "nonphysical matrix-field-effect," as needed by such expanding-evolving mind, to be *pre-existent,* coming before mind, or even evolution, as in a "Cloud Nine" fancy. From this perspective, such field, suitable to be mind or soul's matrix when its body goes, is generally considered to have been created by "superior forces" of some metaphysical nature, laid out for us to discover and occupy if we pursue it arduously enough and deserve it sufficiently. For our purposes, however, mind and the "matrix-field" to which mind must relate to sustain itself are a mirroring phenomenon, each giving rise to the other as a response to evolutionary pressures to move beyond limitation and constraint. While such a Darwinian viewpoint as outlined here does not rule out superior forces pulling strings behind the scene, it makes such hypothetical add-ons simply unnecessary.)

All of this is to suggest that generations of this growing awareness of death, and the continual longing by our species to move beyond it, set up and feed into such a "field" of possibility. The field itself thus grows through a continual reciprocal action between mind and its longing to realize—make real—its imagery for moving beyond its own boundaries.

Anything capable of being believed is an image of truth . . .

WILLIAM BLAKE

Marghanita Laski, in her book *Ecstasy*,[2] investigated the phenomenon referred to as the Eureka! experience, occurring liberally in the physical sciences, the arts, and religious-philosophical pursuits. The Eureka! experience suggests that if mind centers on some "image of truth" with sufficiently passionate longing, attention, and perseverance, the mirroring effect of creation can and may bring the empty category imaged into a manifestation or concretizing of that image, contributing to the "content" of such field of potential itself. This, as in all evolutionary processes, is subject to the stochastic-selective element underlying all emerging process.

Laski documents Sir William Hamilton's 1843 discovery of his famous quaternion theory, a precursor of modern vector analysis that demonstrated the mathematical argument for a fourth dimension of space (thus the term *quaternion*). Hamilton had conceived of such a possibility as a quaternion function in math and had spent years trying, with no success, to solve the particular mathematical enigma involved. His wife reported that, time and again, he would grow discouraged, vowing to pursue the matter no further, only to return to it from a new angle.

Finally, after fifteen years of roller-coaster pursuit, Hamilton seemed to have truly quit, vowing to waste no more of his life on his passion. Shortly thereafter, he asked his wife to accompany him on a walk to a meeting at the Royal Irish Academy. As they crossed a little footbridge into Dublin, Hamilton's mind no longer chewing away over the quaternion enigma and finally at peace, the answer arrived. It came as a flash of insight, the entire theory presented in highly symbolic fashion in a split-second, "out of mind"—whereupon he carved the preliminary formula into the very stone of the bridge upon which he and his wife were walking. Hamilton reported that he knew in that moment of "absent mindedness" the complexity of the insight given him was such that another fifteen years might be needed to translate that initial symbolic web into fully rationalized mathematical terms.

August Kekulé, the famous Belgian chemist, had worked at great

length on the problem of chemical structure, a problem he was determined to resolve, to no avail. One day, wearied with his efforts and frustration, he retired for a bit of nap in his easy chair at the fireplace. As
he drifted into a reverie there appeared directly in front of him a clear
image of a snake with its tail in its mouth, forming a peculiar circle. It
flashed into view and disappeared as quickly, but Kekulé saw in this
strange image the long-sought answer to his problem. A lengthy struggle was required, however, to translate this answer into the necessary
chemical language. Out of this struggle emerged his famous theory of
the Benzene Ring, foundation of modern chemistry.

At a reception given in his honor by the scientific community,
Kekulé was questioned concerning how such a complex, esoteric, and
uniquely original discovery had occurred. "Gentlemen," he responded,
"we should dream more often." Kekulé's famous response has brought
many a challenge from hard-line materialists and Neo-Darwinists who
try to debunk any suggestion of a "psychic-spiritual" or nonsubstantive
element involved in "true science"—as the Benzene Ring surely was.

Laski's Formula

Laski's list of such events is long and rich, particularly when we consider the religious, philosophic, and esoteric elements she detailed at
length. This Eureka phenomenon involves creation itself, endlessly
open, beyond any specific type of creative result involved. We tend to
heed only those creations which are concrete, visible, useful, and commonly shared, but the *process-function* involved has no boundaries, limitations, or selective requirements. In summary form, Laski's outline for
this process is as follows. First, there is the passionate pursuit in asking
the question—staking out the end-goal of some burning issue or desire
that a person determines to accomplish, discover or experience, and sets
out with will and determination to bring about at all costs.

Second comes a "gathering of the materials" as assumed will be
needed by the answer or conclusion, and which generally have some

resonance with the nature of the topic or end-goal. Such pursuit takes place on many levels, consciously or unconsciously, and may go on for years, with an exhilarating sense of discovery along the way. Third comes a "plateau" period, wherein all possibilities seem exhausted, leaving a period of stagnation. The luster of the unknown dimmed and dull, most people tend to quit at this point. One may, however, suddenly think of some angle left unexplored, some possibility that might yet yield the treasure, and off we go again. Fourth comes the dark night itself. The seeker "bottoms out" and truly quits. At this point, the goal no longer entertained as even a possibility, the answer may arrive—or still may not, this being a stochastic realm. Idling about, thinking of nothing, the answer may arrive unbidden, filling the vacuum of thought in a single flash—an answer, which may have never been seen by anyone before, since not existent until "translated" into the common domain.

And here, in this casual mention of "translation," we find both the fifth step in Laski's formula and yet another hurdle to our common logic. Always, in a true Eureka experience of this sort, bringing something new into our world, the answer arrives in symbolic form, which must be translated into the common domain, or the language of that discipline related to the asking of the question. It never arrives "spelled out" in laborious fashion. Nonetheless, mathematical breakthroughs come to mathematicians, since only they have the background to ask the triggering question, to recognize the significance of the symbols involved when the answer arrives, and the materials and strength of mind needed to translate those symbols so that others might share in them. (To me, Kekulé's snake would be a sign only of hallucination or alcohol.)

Moments of Emptied Mind

The recipients of such enlightening revelations insist, however, that they were not thinking of the issue—or thinking of anything at all—when the answer arrived; all claim that the answer arrived in a single flash of insight, never "piecemeal," was highly symbolic in nature, requiring strenuous and often very lengthy translation to put it into the language of the discipline involved, and that they, the recipient, had nothing to do with the answer itself or its arrival. All attest that the answer arriving was radically different from any of their expectations and was purely gratuitous—a gift given.

In each of the cases known, this recipient's claim to a gratuitous gift-given has brought a chorus of vigorous denial from the academic-scientific world right down to our day, in spite of the high-caliber minds involved, and despite the stunning greatness and novelty of the discoveries made. The receiver of the Eureka! is doubted concerning his own experience, and all but called a liar or self-deluded by the skeptics. To academic science a Eureka! could only be a discovery of a pre-existent hard-core fact in the material world-out-there, if it is applicable to science and thus valid. That "mind" is involved in a creative movement over and above mind itself is just not acceptable, while ironically most major sciences themselves rest on and arise from a number of related Eureka breakthroughs.

Equally ignored by conventional thought is the odd fact that such Eureka answers are neither a composite nor a synthesis of all the "materials" gathered in the quest for that answer. The answer is generally at a radical discontinuity with the known materials of the discipline involved, and equally different from any of the discoveries or "facts" gathered by the recipient along the way. One wonders, that being the case, why the long search for the answer was necessary, if none of its "gatherings" had any relation to the final result. And, for that matter, how the "facts" gathered, though not part of the answer given, were nevertheless recognized as significant to the general realm of that answer, or were pointers toward it.

And in that question lies another of these valuable insights to be gained from this phenomenon: the issue of resonance—that major function by which the brain works and fields form. The brain-mind involved must have a rich, resonant field into which the answer can arrive and be tended, rather as well-tended soil for a seed given to flourish, even though all ordinary brain activities must be suspended at the point of Eureka!-arrival itself, leaving only that resonance prevailing.

Fields as Creative Process

Fields of potential are not only active forces or intelligences within their own domain, they are also creative forces. Why would the mind of a Eureka "recipient" have to be suspended or empty for its own answer to present itself? And why would the answer have to be so obscure and symbolic? It seems the mind, commonly defined as personal, was not in and of itself the cauldron of creativity forging the revelation. The recipients' insistence that they had nothing to do with the revelation or its arrival—common to all such Eureka reports—need not be explained away or attributed to false modesty. Though this is surely true at the individual or personal level of awareness, the whole event is a cosmolog-ical-ontological function well beyond the personal, even as it embraces all minds and much more.

In each case, the personal field of mind fed into a commonly held or universal field of like order, one over and above any individual mind-brain, yet obviously in some sense the product of that individual mind and others in the field. The very fields of mathematics (in Hamilton's case) or chemistry (in Kekulé's case) are themselves the cumulative result of the work of all mathematicians and biologists now and in the past. We create fields even as we interact with fields. We speak of going into the field of medicine or the field of architecture or the field of engi-neering, but field in this academic regard concerns shared social activi-ties, and denotes an aggregate of intelligent energy and potential.

Such a field as experienced in Eureka! events has no inherent local-ization; it localizes according to whatever mind is interacting with it. There is no field of medicine without doctors, while no doctor is that field. All is reciprocal. Field-effect is universal and personal. It is a pro-cess, a dynamic, generative aggregate of intelligent potential. Fields of knowledge such as mathematics, physics, music, and so on are in a continual flux of arrangement and rearrangement brought about by the constant input of materials from people studying and employing the fields as well as those unconsciously interacting with them. In the case of chemistry, all those generations of students, professionals, amateurs, career scientists, and lonely thinkers, mulling over the mystery of chemical structure, fed into the field that led to Kekulé's Eureka! experience. There is only reciprocal action between mind or minds and fields.

Because of this, any field contributor might automatically be in line as a possible target for receiving some symbolic answer or Eureka! discovery brought about by and within that field. The one stipulation seems to be that an individual mind must be idle, vacant, or inactive at the precise instant of the field's creative action. Out of the ferment of a field of potential interacting with many individual minds, one of those minds may, by chance, be struck by the lightning generated in that particular field. Without a resonant mind to receive a field's creative invention of the moment, nothing could happen in either field or mind. Creativity lies in the reciprocal relation between individual and field. (Here we find an example of Darwin's random mutation and selectivity in its truest cosmic application.)

When Lightning Strikes

Electricity generates in cloud activity and gathers into aggregates of electrical force, moving from cloud to cloud, acting as attractor in each, until a saturation point is reached in such an aggregate. This collection then literally "seeks out" a like resonance on the surface of the Earth below.

Meanwhile, on that Earth surface, electricity gathers at specific points, and likewise moves through resonance to attract and gather together with similar electrical charges in that local terrain. You may feel your hair tingling should such a surge move through your terrain, gathering forces so to speak. When that gathering together is of sufficient size and power, it seeks out a resonant energy of like order in the sky above.

When that aggregate in the clouds moves into close enough proximity to the Earth charge for the resonances to be sensed, the Earth-aggregate, which is far smaller and less powerful, gathers its entire force and, seeking out the highest point of the surrounding terrain (hopefully not some upright body such as yours or mine), literally "leaps up" to attract the far greater and more powerful electrical cloud above. The greater charge above then fuses with the ground charge and follows it down, walloping full force the point from which that weaker ground charge made its bid-for-union.

We might note how this scientific account (drawn from *National Geographic*) seems to make near-sentient characters of a supposedly non-sentient force, which should be a strong pointer toward our previous discussion of the related series of holonomic torus fields of heart fusing with Earth and thus with Sun. Our current notions of sentiency may be a bit too localized within our own personal frame, and may in actuality be more universal than accepted heretofore.

"Mechanical Excellence Is the Vehicle of Genius," claimed William Blake

As Found in Mozart's "Round Volume of Sound"

In Marcia Davenport's biography of Mozart, she relates that in his late mature stage of composition, a commission for a new work of music—such as a symphony, concerto, or quartet—might be given him at a time he was too busy to attend it. Mozart would put the new project on hold; tuck it away in the back of his mind, so to speak, in order to concentrate his energies on his present works. Often, however, the new work commissioned and "put on hold" would, in some odd moment, break into his awareness right in front of him, unbidden, as a visual-auditory "round volume of sound" (reportedly Mozart's words).

This "round volume" contained, in its instant of appearance, the entire proposed work in its completeness—every sound, phrase, section, nuance, shading, dynamic, and instrumentation present in a pristine perfection. And all in that single instant that yet could spell out in linear time the auditory manifestation. When he spoke of this happening more often as he matured, his account led to the myth that attributed his genius to his being but an amanuensis of the Muse, who took no part in the actual creation of such works other than this "secretarial role." "But nobody knows," Mozart lamented, "how hard I have to work" to translate this instant flash into the torrent of notes that must be spelled out in ink on paper in order for the "round volume" to be played by others and take its part in the musical world. This translation was no easier or harder than ordinary composing, both of which required the labor of getting such expression from mind to pen and paper. Both were processes of the same mind, and a clear example of Mozart's "mechanical excellence," ever ready to serve his genius when it came.

The similarity of this with other Eureka experiences is obvious. These are field-effects residing, we might say, in that field of all fields, the "akashic record" or "astral realm" (as you please). But bear in mind that the snake of Kekulé's vision became an integral part of our very concrete scientific-technological world, and each of Mozart's hard-earned Eurekas is another masterpiece in music's rich history.

As a final clincher and bonus here, this Mozart field-and-mind process took place when I was in graduate school, through a professor of musical esthetics (whose course was a fascinating journey through various branches of philosophy). This professor was also an active concert pianist, frequently away in travel-concert performances. He spoke to us of a particular Mozart sonata he considered the most esthetically perfect artwork ever created, and easily his favorite of all music, in all forms.

For one of his concerts he had chosen this sonata for his presentation, feeling, he said, near-humility at being the instrument of its re-creation. Just before beginning to play, he leaned back for a moment to immerse himself fully in the work, his love for it filling him. And at that moment there appeared right before him a "round volume of sound"—his own words—with every note and phrase in its most pristine purity and power, the sonata sounding in its linear time, all in a fraction of a second. This was, he said to us students, the greatest moment of his life, its summation and highest point. At which point he began the performance, which proved to be the greatest playing of his life, a point of perfection he had never known nor quite gained again. (He had not read Marcia Davenport nor heard of Mozart's account.)

My point here is simple: that creation of a sonata, originally an interaction between Mozart and the field of music, was permanently within that non–temporal-spatial field from that point forward, ever ready, perhaps, to be activated by a person with the right resonance, clarity, and ability to be absorbed into that "field of reception." And— no small item—able to be "breathed by it" and thus "translate" a potential state into actuality. Mozart's "round volume" was Mozart's own genius expressing itself, his "mechanical perfection" having provided for and then making way for that genius. When my professor activated Mozart's sonata, breathing it into life again, he may well have breathed the spirit of Mozart himself back into our world-awareness for and as that episode. And how do we know but that in every replay of one of Mozart's great gifts to us, he himself takes part? (I must confess that having attended Robert Sardello's remarkable course called "Caritos, the Honoring and Care of the Dead," this notion, which had long lay in the back of my own mind, seemed concretized and validated.)

Innate Field Resonance in Childhood

In the mid-1970s, Harvard University's Burton White published a milestone in child development, *The First Three Years*. He and his staff had found a scant 3 percent of American children brilliant and happy. This fortunate 3 percent came from a variety of social-racial backgrounds, but were noted to have one common, exceptional characteristic: *these children spent an inordinate amount of time in blank, open-eyed staring, doing nothing at all.* Anyone bothering to look squarely into those open eyes might declare "nobody is home." Today, any number of dysfunc- tional labels might be laid on such a child, with various modifications of such behavior undertaken, from therapy and drugs to coercion and punishment. Fortunately, as it was, they were left alone.

Jean Piaget, the Swiss biologist turned child psychologist, referred to children in this early pre-logical state as "children of the dream." Given nurturing, security, and the privilege of being let alone, their main occupation was daydreaming and playing, lots of each producing brilliance and happiness thereafter.

Rudolf Steiner referred to three- to seven-year-olds as "etheric chil- dren," one foot in the world shared with us, the other in the etheric world. (Eastern philosophy speaks of the subtle realm, while physicist David Bohm might refer to the "implicate order.") Around age seven, these etheric children, ready or not, undergo a rude shifting into the "real world" of us logical adults, their inner world and its creative imagination generally—but not always—lost to a playless world of grim necessity, schooling, and full enculturation.

Consider how anything similar to this open-eyed staring trait in children seriously disturbs today's parents, caretakers, teachers, and the culture at large, who consider such "withdrawn" behavior pathologi- cal, perverse, or anti-social. "Don't just stand there, *do* something!!" A great treasure is at stake in these pre-logical years, generally lost in such restrictive constraints and arbitrary "doing something" to conform to a destructive cultural pattern.

Field Resonance in Indigenous Cultures

I have known two Australians (European descended and educated)— one a medical doctor, the other a university professor—who were adopted into one of the few surviving Aboriginal groups still living in traditional ways. The university lady remained eight years living in "walkabout" with her totem family, into which she had been adopted (or "married," as that family spoke of it, not quite the sexual union we might envision). From our long conversations I got at least a dim grasp of the great gap between the lived states of the two cultures, Aboriginal and white Australian.

She claimed that none of our Western anthropologists studying the Australian Aborigines (not even Levi-Strauss or Laurens van der Post, whose works I had long admired) had any vague idea of what the real world of those native people was, for such a world was simply closed to our European senses-five; it takes some strenuous "doing" for Westerners to drop enough of their own mind-set to hook up with, or open to, that other-perceptual world. Nonetheless, there apparently have been a number of real cases of Western minds opening to the Aboriginal Dreamtime, although the people involved are generally quiet about it, since the gap between the worlds is too great to span through explanation. Personal exploration is essential in such circumstances, and I envy those who have had it.

For our current purposes, there are perhaps a few tentative observations we can make. Through a rigorous upbringing and training in a practice followed for hundreds, perhaps thousands, of generations, the Aborigines have built up a powerful field-effect. Each initiate living the Dreamtime contributes to and strengthens that field as they live it. Traditional Aborigines have developed a balanced intelligence, which moves naturally between their Dreamtime (akin to Steiner's "etheric-dreaming"?) and the practicalities of everyday living.

As with the brilliant-happy three-percent Burton White found, some of the Aborigines apparently spend a great deal of time in open-eyed, blank staring, so oblivious to the outside world that flies crawl unmolested across their open eyes. To an observer, it can seems that "nobody is home" there either, at such times, which may give us some hint of why our children drift into realms of reverie on some level, given the chance. We have no idea what the rewards of such reverie might be for an adult trained in such ways from birth, and probably never will know. As Robert Wolff laments in his book *Original Wisdom,* concerning his many years of interaction with aboriginal people of the Malaysian rain forest, we have no idea what we have lost.

Surely the Australian Aborigines are but one of a host of global indigenous cultures whose "original wisdom" and access to intelligent mind-fields is either already lost or severely at risk. The question for us becomes: Is it too late to recognize, in nonromantic fashion, the value of such original wisdom? Few Westerners will have the opportunity to formally participate in the "old ways," but serious reflection on the deeper meaning of these ways may play a role in the evolution of mind as we are speaking of it here.

The Double-Edged Sword

In sharp contrast to mind-fields accessed in the "old ways" are many of the charged and potent fields generated in our contemporary milieu. Darwin once wrote that any action repeated long enough can become a habit, and any habit repeated long enough can become the equivalent of a genetic trait or inherent field-effect within us. Howard Gardner's proposal of our major intelligences having become generic and inheritable may well be laid at the feet of this tendency defined by Darwin. Thus the possibility of a current "fad" becoming a cultural habit and possibly, over time, a species-wide field-effect picked up and carried genetically, is but an extension of this tendency.

Far more sobering is the obvious conclusion that negative "fields" of emotion sweeping a people (as discussed in previous chapters), gaining momentum and strength generation by generation, might well become repetitive and finally inherent within a cultural pattern of behavior. Inherent intelligences and abilities active from birth might well need but a bare trace of some habitual-become-potential-trait for such a trace to act as an attractor, through which these field-effects fill in related content. Through ordinary reciprocal functions between individual and society, such a trace would grow into a full function as we grow and mature.

All such functions reflect field-effects inherent within our structures of knowledge and resulting worldviews, whether they be positive and life-affirming or negative and destructive. Such field-effects can enter into creation itself, and show that to some unknowable and immeasurable extent we are an inherent part of the reciprocal action of creator and creation. As we "loose on Earth we loose in heaven"—which may or may not then "rain on just and unjust alike."

Questing Creates Its Answering

And so we come full circle, faced not only with the inevitability of personal death, but also with the looming prospect of death on a much larger scale. What we are suggesting here is that human life, confronted with the apparently insoluble problem of death early on, nevertheless, in its striving to transcend or "move beyond" even this constraint, has now brought about the necessary *potentials* for achieving such goal, although hardly in a direct, miraculous wave-of-wand, but rather, indirectly through natural process. In typical strange-loop form, the question— ever more pressing in light of our pan-global destructive potential—is entering into the creating of its answer, with the answer entering into and continually clarifying the nature of the passionate quest itself, mirror to mirror. In this way self as brain-mind has entered into the evolution of self-as-heart.

Matrix Shifts

Thus, we can see the preliminary outlines of an evolving shift of matrices from tangible body-brain-heart concreteness to the nontangible "abstractions"* of a non–temporal-spatial field of mind. Bear in mind, however, that our original matrix of tangible primary survival systems, instinctual and strongly entrenched in every cell of our body, is always

---

*Here, "abstraction" is not meant as philosophical or theoretical "flight of fancy," but rather as an actual movement into a nonphysical domain. We could equally say "extraction."

in this body as part of its maintenance system, and will be until death-do-us-part. This new "non–temporal-spatial state" takes time and extensive development to establish, not being an overnight quickie operation.

Generally this development of a more advanced evolutionary system boils down to a conflict between that most primary structure of our old "reptilian" hind-brain and the open-ended potentials within our latest evolutionary brain, the prefrontal cortex. And here we arrive at a paradox: *for mind-as-emergent to "escape" and move beyond these ancient survival instincts, it must fully accept and embrace the concept of death itself.*

Everything in our culture is designed to avoid such a head-on collision through hypothetical escape hatches, by which we can maintain our illusions of bodily immortality. The issue we face is going beyond instinct (the fear of death), *and then moving into that very unknown state of mind our instinct guards against.* Such movement is the full and nonnegotiable acceptance of death, without hidden aces up the sleeve as sold by culture. To give up the notion of survival is equivalent to giving up one's life for greater life, as a wise man once put it, and such an illogical and contradictory notion proves to be a biological observation, not a religious one.

Further, as a cautionary note here, in this emergent process of mind will be found those same ironclad issues of stochastic profusion and narrow selectivity found throughout evolution. That same wise sage observed that "Many are called but few chosen," which involves far more ontological aspects of selectivity than religious aspects. This stochastic aspect centers around mind becoming fully emergent from its original matrix—a process dependent on having created a sufficient matrix to move into—as in Laski's Eureka events where a matrix itself forms by the nature of our question and movement toward its answer.

Only a fully developed mind can contribute to and so become, in its turn, the matrix for further evolutionary states. Emerging out of a fully emergent process itself, like an abstraction—or "extraction"—out of abstractions, mind can then move beyond any and all known physical functions and their restrictions. This, of logical necessity, includes the phenomenon of death, wherein further and even more complex movements extracting or abstracting out of abstractions are involved.

When poet Blake said that "Anything capable of being believed is an image of truth," he did not imply that some belief is the truth as such, or even necessarily "true." But the capacity to create such images and even believe in them is the way we are "made in the image of God," even, as with any strange loop, the way we make God in our own image, both images being necessary. In such imaging, being the way by which all our vast creativity unfolds, some creative efforts indeed become true, even on a broad consensual level. Imagination—creating images of possibility—is the stuff of life and its creative evolution, and part of what our human story is all about. (Our failure, therefore, to foster imagination in children is a fatal error for humanity as a whole, threatening to undermine the very creative potentials we are exploring here.)

Evolution's Work-in-Progress

So this species-wide development of a mind capable of recognizing personal death and then going beyond it has taken time to establish as a possible field of potential. Mind, as self-awareness, must now further develop the capacity to imagine and project beyond its present physical "embeddedment." If the resulting imaginal state is entertained over time, it can set up a strange loop, mirroring between image and its object, beginning early on, which is in process today. The issue lies in creating a matrix stable enough for the mind to reciprocally interact with and achieve a stable-state within itself, which is rather like the stability of heart's torus—such stabilization also being an intentional-imaginal process. This stabilization is of course one particular aspect of the imaginal dynamic involved in grasping and realizing the related series of holonomic torus fields of heart fusing with Earth and thus with Sun. In such imaginal activity we may discover that our notions of sentiency are a bit too localized within our own personal frame of reference, opening the prospect of a more universal sentience than accepted heretofore. In this we have, as well, a near perfect example of what Rudolf Steiner claimed was our mind's role with heart, and heart's "next level of evolution."

Finally here, I would first call attention to the overall capacities activated and exercised through Robert Sardello's and Cheryl Sanders-Sardello's School of Contemplative and Spiritual Psychology, and, second, I would again call attention to the sober fact that Nature operates by profusion and selectivity, and of necessity the selectivity grows more stringent the more complex the process involved. Constraint and limitation, like the horizon, always lie right beyond. So, that "many are called but few chosen" is not a dictate from some hell-dealing judge or loving arbiter on Cloud Nine, but simply the way this creative cosmos is set up.

TEN

MIND, SPIRIT, AND CREATIVE FIELDS

This book proposes a process which gives rise to our universe and everything in it instant by instant, wherein nothing that happens or conceivably can happen is other than process itself. "It" isn't *process* that creates—"It" is creation. We cannot, in actuality, freeze or segment the process of creation (a verb signifying movement) and make of it some sort of nonmoving noun-thing that is fixed. Process is "itself" simply a product of the movement, which "it" is—a rather tautological mirroring or strange loop. Thus again, as our wise man of a couple of millennia back said, the human has "no place to lay his head" as the animals have, since the human's nature is of that Spirit-process itself, eternally fluid and moving, ever new and recreating itself.

To consider process as a thing-object-phenomena existing outside the creative process as something in itself, is, as Alfred North Whitehead pointed out, an "error of misplaced concreteness." We tend to fall into this error repeatedly, referring to a product of thought as an objective thing-out-there. We take fluid events, movements, actions, or functions and treat them as static things or fixed-objective forms that we can then manipulate. Physicist-philosopher David Bohm echoed Whitehead's concern, regularly pointing to the serious implications of reification—treating an abstraction of mind as substantive, final, and permanent.

In our further exploration of the potential of mind-fields—and even more so as we begin to consider "Spirit"—we must be aware of our inclination to slip into reification and misplaced concreteness, and to then judge events or ideas from these fixed criteria. At the same time, we need to recognize that belief, imagination, and mythological overlay—which can lock us into an inflexible mind-set of reifications—equally play a critical part in our evolution of mind. If we are alert, this interplay between reification and imagination will keep us on the horns of a number of healthy dilemmas.

Field Resonance in Time

Religions, adhered to and practiced by enough people over enough time, can become field-effects and to that extent valid from a functional, pragmatic standpoint, regardless of inner or logical coherence. Further, if such field produces according to its precepts, bringing effective change in behavior or thought or both, it is even more functionally valid. Applying historical-materialistic criteria as a final arbiter of "truth," however, is shortsighted.

As the poet William Blake said, "Anything capable of being believed is an image of truth." To Blake, as with that intuitive sage Jesus, "truth" was a pragmatic result, not mind-wandering abstractions. Field-effects, be they cosmological or ideological, can become active structures of knowledge and function as readily as any part of a mind-set, such as our sensory system's response to environment. To live in Spirit and truth is to translate grace through one's actions in gravity.

"If you would possess a virtue you must first assume it," Shakespeare pointed out. Planting a seed of potential into the "field of mind," even if a "let's pretend" virtue only imagined, may attract additional precepts or concepts and expand the whole structure, all beneath our awareness. ("Much water runneth by the mill that the miller knoweth not of," can be read as the good Bard's metaphor of mind's surface awareness floating atop the vast labyrinths of activity within body-brain-heart's creative interactions.) Thus, we might well begin to "unconsciously" act out the originally imaged virtue, which indicates that what was imagined has grown until becoming a valid, working frame of reference. Producing products of mind-experience in keeping with the premise of its genesis, again indicates grace preceding gravity, or effect preceding its cause, recognized so clearly by poet Blake.

Modeling

Strange Loops in Time

Nearly every aspect of infant-child development hinges on such play-pretense of the adult models they imitate. The greater the freedom to play and the more real and tangible the adult models, the more complete and effective the final structures of knowledge-ability resulting in that child's life. A virtual reality inserted at any point as a substitute derails the development.

Picking up a spool from Mother's sewing kit, the child's inner memory-imagination might create a powerful truck or automobile projected onto that spool, the child playing with its creation for hours. As Lev Vygotsky spoke of this, the child creates an inner world, which he projects onto his outer world, and plays in the imaginary world of his own creation—a synthesis of the inner and outer imagery. This gives a world over which the child has dominion. He has been God creating a world, and then is the adult in command of that world. Later, in its actual adult form, creating an internal image projected outwardly can concretize—make real in the physical world—creative constructions of the inner world, which might then be beheld, perhaps, as a great discovery awarded the Nobel Prize—again the cosmic play of creation.

Consider how a mythological overlay, as on a great figure of history, may overshadow the original figure overlaid. By the attraction that original figure holds to any related or resonant fields of action or potential, these related fields gather and add to the power and significance of that original figure *and the field-effect it involved,* the attraction to the growing overlay itself growing accordingly.

So the historical-evolutionary value of such mythological overlay is not so much to enhance the validity or ironclad "truth" of the original figure, as to give each current generation exposure to a significant "attractor-of-greatness" (as most overlaid historical figures are). Therein any attracted individual's own latent greatness, finding resonance with that mythical figure, might also find a means for working out, expressing, or translating such hidden greatness itself within him. All this is an extension of the "model" being imperative to the needs of a creative imagination to jump-start a creative action within itself. Such event might then be found "out-there" in the world, making this a Divine Play in poet Blake's sense. (And, again, giving the grounds for our Great Being's "always becoming what we have need of him to be.")

Such a dynamic is a vital part of the model imperative underlying child development. In order to be activated and take part in our reality-making, any inherent intelligence of ours (as detailed, for instance, by Howard Gardner)[1] must be activated by the presence of a living model of that intelligence. Then this reciprocating model-capacity will tend to unfold, flower, and bear fruit in our life. The "model" is both stimulus and mirroring guide of development of that inherent capacity being activated, and through this reciprocal action the model is itself strengthened and expanded thereby.

Every child learning to speak enhances the language-field from which the model the child followed arose. This is another example of the strange-loop mirroring wherein a new creation loops back onto older ones to both stabilize and strengthen the new, which advances and further develops the old and provides a launch-point for further expansion-employments of the original potential (as with brain development, discussed in chapter 2). This evolutionary procedure is one I have long claimed on behalf of the mythical overlays on Jesus: the overlays far outweigh either the value of, or the credibility of, some near-hypothetical originating figure of some hypothetical "real concrete" or "true" history. All such "calendar history" of an original figure turns the entire movement of mind back into the past, replicating that past to some extent, and *subtly shutting off any future-becoming it might hold for us,* which is the real thrust of great beings in history. Thus, we stay locked into a destructive past with its vast injustices becoming ever-present to us. The immediate "now-presence" of that mythical figure in our inner life of mind is its value, one that needs no historical verification. Such verifications as attempted could only be another variant of mythical overlay at best.

This play of imagery between a nonphysical potential, such as one of Howard Gardner's innate intelligences, and the concretizing of it through a living model's physical demonstration, can initiate in the child a "mirroring" of the model and establishing such capacity in the child's development.

Judith von Halle and Field Resonance

This brings us to Judith von Halle, whose personal history proves an enigma that makes for a stumbling block to ordinary logic. She received her doctorate in architecture in the United States and returned to her native Berlin to practice her craft. A follower of Rudolf Steiner, she would surely have developed a mind-set open to Steiner's "higher worlds" (analogous to *mind-fields* as used in this book). Thus, she would be capable of opening to, grasping, and entering into those same phenomena.

These "higher worlds" are sometimes referred to as the "akashic field" in occult terminology, a hypothetical universal record of all events ever occurring. This subtle aggregate of all events and memories brought about by and in human experience is rather a "mirror of the universe of humankind" and all its happenings.

Such a plethora of field-effects, aggregating according to general resonance, can function selectively, manifesting as any intelligence or ability. The akashic field is ostensibly a "list" including everything, from nonsensical imaginative fabrication to atom bombs, lasers, or whatever. Such history does not consist only of facts and figures available to a logical temporal unfolding, but also of any resonances as constitute our brain, its "memories" and workings. What one finds in that field depends on the maneuvers of mind one employs to access it. We seek and we find—according to the procedures used in the seeking.

An akashic record might have no time-space correlates—nor need any—to simply "be there" in its subtle realm. Field-effects may or may not contain specific contents within any aggregation of that field, and no hard and fast conclusions concerning such hypothetical "field of all fields" can be made. For the materials or content such field may contain would likely form through our search itself. Our search may function as a seed of potential forming in that field. Or a search might be initiated and impelled by some form of resonant memory of an event lingering about in our mind, rather like an echo or reverberation. Recall here Paul MacLean's proposal that the brain functions through resonance, not necessarily through facts or information in our ordinary encyclopedic sense.[2]

Judith von Halle may be an example of a field of information-knowledge-memory being enlivened and lived out in a person's own history. Such an information-knowledge field, expanded, enriched, and elaborated on by its reciprocal interaction with a living, intelligent mind, can attract and incorporate other field-effects having resonance with it, the original event thus growing in validity and depth, and becoming ever more powerful and attractive. Gerald Edleman's concept of memory as a re-membering, or gathering together of resonances from a variety of sources into a singular event, lends support to this perspective. Such event, even when logically coherent within its own frame, may or may not be valid in a larger context. Lived out and dwelt on, however, such a re-membering may attract to itself related events or resonances, which bring growth or expansion to the original, and is a way by which related potential can unfold. Every new relating resonance loops back onto the original possibility, expanding and enhancing its attractiveness and bringing such expansion so long as the re-membering remains active.

Von Halle's experience (as she personally has detailed in an intriguing biographical history)[3] tangled me in a labyrinth of associated phenomena, forcing me to abandon some of my pet prejudices and judgments. Judith presented an enigma so niggling to my mind-set I could not let it alone, nor fully cope with it.

The most notable feature and apparent inception of von Halle's spiritual adventure was experiencing the stigmata—the sudden appearance on her body of the wounds received in the Crucifixion. This phenomenon itself has a long history in the West, generally among those of the Roman Catholic tradition. It runs from Francis of Assisi, considered the first known historical figure to experience the stigmata (and who may have thus introduced such a notion into that field-of-belief as an integral part of it), to the Catholic nun Therese Neumann, who died in 1962, whose experience was thoroughly examined and authenticated on every level—medical, scientific, and so on. The stigmata phenomenon traditionally appears on Good Friday, lasts the expected three days, leaving at Easter, mission accomplished. Therese Neumann experienced the stigmata at Eastertide all her life, and lived for many years without any food other than the single wafer given at daily Mass, a paper-thin wafer as near to nothing as can be made.

Von Halle's wounds appeared spontaneously. They bled, as appropriate to tradition, but did not go away. Instead they were accompanied by an ongoing, vivid, full-sensory reenactment-incarnation of the entire Crucifixion event itself. Later this ongoing episode was apparently influenced by—and shaped within a frame of mind unique to and an enlargement-refinement of—Rudolf Steiner's vast creative vision of the "Mystery of Golgotha." And while the unfolding of Steiner's cosmic drama within Judith's own creative mind shows how mythical overlay can build and take on ever-greater potential for the mind's creations, it proves a stumble to our acceptable academic or even occult patterns of conscious awareness.

Judith found herself not just witnessing, but equally living out the

Crucifixion, both witness and victim, so to speak, rather reminding one of Wilder Penfield's patients perceiving two reality-events simultaneously. Judith's ongoing experiences took place, however, through a different modality of mind than most nonordinary phenomena. Robert Sardello suggests her mental capacity embraced an ethereal, perhaps a-causal realm of "clairvoyant consciousness," giving Judith an immediate present-time experience of the Crucifixion drama. The proposal is that hers was not a memory recreating an original event, but a temporal shift of her sensory-system and mind-awareness into the time frame of the event itself. She was not just witnessing, but an integral part of that event as itself, in its own actual time-frame—a sensory warp-in-time, so to speak. This gave her a present-time presence in the whole drama. Once established in her, this clairvoyance took over and developed its own ongoing frame of reference with clarity and logical cohesion.

Here again, we have the power of a suggestion to become a nucleus of a resulting experience relating to the present moment—a memory-event of a past moment. Bear in mind that such akashic-record-frame events cluster according to resonance, not according to a logical time sequence or spatial proximity.*

At any rate, Judith's ongoing experience grew in scope and vastness, replicating in her (perhaps ethereal or "phantom") body all the agonies originally experienced or imagined to have been experienced in Judith's own visionary synthesis. More and more details came to her notice, her story growing in depth and power over time, as an event she witnessed

---

*Temporal-spatial measurements are not part of the memory-fields in which we can take part in our all-too-real time-space, but crossovers between such "event boundaries" may well be a key element within the strange-loop phenomenon. "Subtle body" phenomena play as vigorous a part in Eastern ontology as physical, and one can't hang around and take part in a vigorous ashram life in India without encountering one's subtle body in a variety of ways. That there is a crossover between subtle and physical is obvious, and logically necessary to explain what goes on and happens to oneself. Subtle effects can include spiritual or psychic effects as determined by one's own definition and logical accounting, which accounting can build one's comprehension, opening one to ever-greater dimensions of such an experience.

and experienced. Further, her awareness apparently expanded over time to encompass more and more of the vast historic, theological, and intellectual ramifications involved in that event, cast within what may be a prime example of Blake's "divine imagination."

That this presence-experience of the Crucifixion did not leave Judith, but unfurled in her life day by day with ever-enlarged ramifications, is exemplary of the mind's creative capacity. Writers will attest that the longer they live with, ponder, and conjecture on some character or story line they have created, the greater the dimensions their writing on the issue grow—the characters taking on ever-greater validity and realness to them, all but dictating their own destinies. To denigrate either psychic or spiritual experiences on behalf of ironclad, scientifically verifiable facts as the only "real" simply isolates the mind into its own cavern. The list of those who expand the human mind—von Halle, Bernadette Roberts, William Blake, Meister Eckhart, Goethe, on and on—has been and always will be endless. Such experience points up the dimensions of our creativity behind the scenes at all times, in all we do or experience.

Transcending Physical Constraints via Field-Effects

Of equally serious significance was von Halle's body immediately rejecting any form of food on receiving the stigmata. In the ensuing years she has been *unable* to eat, her body rejecting food as though extremely allergic to it, the very idea of eating nauseous to her. Meanwhile her ever-expanding vision-sensing of the Crucifixion continues in an ongoing event engulfing her life, taking on ever-greater dimensions, all finding an orderly, logical premise within Rudolf Steiner's intensely intellectual-creative framework. Steiner's well-known frame of reference, studies, and vast writings have been discussed by countless students and followers for a century now, building a continually strengthened resonant field. Judith's experience may well have had its genesis within that rich and ever-growing frame, but she no doubt moved beyond the boundaries of Steiner's creation, and her own. The two may have given rise to each other in a strange loop—a richly creative-imaginative play of mind through which any participant in the original, including the very one *crucified,* may have found new voice and expression (again, "always becoming as we have need of . . .").

That von Halle's inability to tolerate food seems foundational to her continually expanding Crucifixion experience brings in another aspect of relative and resonant field-effects. In contrast with von Halle, Therese Neumann suspended all nourishment before she began to experience the stigmata. And hardly insignificant are the rich examples of fasting we have of the original mythical-historical figure of Jesus, perhaps a nucleus bringing all this about. In one episode Jesus reportedly had declined his comrades' invitation to dine with them, saying that he had "food they knew not of." Other accounts report his forty days in the wilderness without food, which in itself has copious symbolic meanings and perhaps indicates that such capacities are available to all of us, if unbeknownst and as yet undeveloped. The accounts of Jesus suggest that fasting and prayer can combine to make possible miracles or "interventions in the ontological constructs" (as Mircea Eliade spoke of such) of our world.

Although we have these potent accounts of Jesus fasting, including his forty-day marathon, the ontological significance of the capacity has long since been lost in the labyrinths of religious trappings devoted to proving Jesus's divinity, rather than his role as an exemplary model of our human potential, in which we can find ever-greater freedom. Perhaps the seed of such ontological possibility has again been planted into our human psychic makeup, this time in the general cast of a semi-scientific-occultic frame, and again as a pointer toward an evolutionary movement of mind beyond limitation and constraint, as we have been considering throughout this book.

In a similar vein, we should consider the experience of Michael Werner. A former university professor and currently CEO of a large chemical firm in Switzerland, Dr. Werner's is a fully credible academic-scientific account of his own well-established life without food, through following a specific rigorous process.[4] He initiated his process in 2001 with the "21-Day Fast," a method pioneered in Australia and since practiced successfully around the world. This rather radical twenty-one-day procedure involves a complete and total fast from either food or liquid for the first seven days, followed by two additional weeks of no food, but small amounts of water with some fruit flavoring if desired.

Werner acknowledges that the early phases of the fast can be challenging, but claims that once he passed the twenty-one-day threshold it became easier, and he has since had *no need to eat at all:* "I feel healthier and more vital than ever. My powers of resistance and regeneration are stronger. I'm hardly ever ill any more. Psychologically, I feel stable and mentally enriched, have much better concentration and memory than I used to, and now only need five or six hours' sleep, rather than the eight or nine I used to."[5]

The originators of this process claim to have received the instructions for the fast through clairvoyant means, by which they contacted an intelligence from "the other side." If the participant-candidate holds in mind that the "source on the other side" has promised to assist any candidate as needed, such help will be forthcoming. The needed nourishment from this psychic plane then forms and carries the threatened physical system through those first critical days and on to success. After twenty-one days the initiate can, if successful and holding to the end, either resume eating or elect never having to eat again.

There are two personal acquaintances of mine who, having gone through this process, represent both spectrums of success: one hasn't eaten in ten years or so, is doing fine, loves the freedom achieved, and has no interest in resuming the natural cycles. The other acquaintance didn't eat for eighteen months but got bored with it, missing his meals with friends and such, so eventually going back to eating. He had no adjustment problems, his body immediately assimilating food in proper fashion when called on to do so. Either way the participant goes, the event brings some degree of metanoia, or transformation of the mind.

Thereafter, one simply looks on life from a different perspective than from the grim survival-orientation shaping us ordinarily.

The promised assistance from this phantom "other side" apparently has come through time and again, demonstrating an interaction or "crossover" between states of consciousness—physical, spiritual, psychic, clairvoyant, or whatever—even if such states cannot be fully articulated.* And this crossover of psychic and physical can itself be a serious breakthrough, giving us a new understanding of ourselves, should we pay attention.

Michael Werner provides us with a solidly reputable basis for the phenomenon of transcending scientifically defined physical constraints, yet people continue to ridicule even the mention of such possibilities. The reason for this ridicule and strong tendency to dismiss the phenomenon entirely is simple. As Susanne Langer said, our greatest fear is of "collapse into chaos should our ideation fail us." That people are discovering that they can live without food or nourishment is a potent threat to all aspects of enculturation and its ancient ideations. As Judith von Halle points out, our entire modern world and mind-set would be undermined were we to fully acknowledge that our actual mind-brain-body system can be sustained through spiritually based action. Such understanding could blow our whole modern house of cards away, revealing knowledge of who and what we really are.

Crossovers and Spirit Infusion

"Crossovers" between our known physical-mental realm and the more subtle realm of mind-fields are, I would propose, fairly common. Many

---

*For quite a while in the early 1950s, I took part in the rather rigorous exercise-challenge of "mocking up" a phantom-body while in full conscious awareness of my ordinary physical state. This led to surprising results, and twice I have fallen back on that odd capacity, wherein this "mocking up" bailed me out of a precarious physical predicament. All of which suggests how narrow is the cave of mind we ordinarily accept as all we have available, while real riches lie beyond.

examples we have given here, from the Eureka moment to the phenomenon of the twenty-one-day fast, are of this order. In the case of the latter, the subtle realm can then support the physical, so to speak, in a crossover of energy and sustaining power. On the other hand, there is no "crossover" of Spirit and physical realms, no bridging the event-boundaries of two different realms as found with psychic and physical interplays. *Spirit has no "realm" or event-boundary.* Judith von Halle's sudden body rejection of food was not a "psychic invasion" or takeover of the physical, or substitute of one biological action with another, but one of the many manifestations of Spirit's *infusion* of von Halle's whole being. Spirit's infusion of a person is a different ball game from psychic seizure, channeling, crossovers, or intervention in ordinary causality.

Thus, there is a distinct, perhaps profound, difference between Werner's success with the twenty-one-day process and the infusion of Spirit that von Halle underwent. Werner's experience perhaps indicates a functional, working relation between the psychic and physical realms, whereas von Halle's infusion was an *equating* of Spirit and von Halle's whole being—Spirit, body, and mind. Sameness and resonant similarity are not equivalents. Regardless of all these distinctions, the cultural counterfeit under which we have labored these many millennia is not about to let some "mind-over-matter" or psychic-spiritual awareness take serious root. This conflict will go on within each of us simply because of the hind-brain/fore-brain activities still prevalent within most of our makeup.

Perhaps a larger frame of reference and identity for us could arise from a link of this Australian-based psychic-spiritual locus and Judith von Halle's Steiner-based experience, so long as we remember the distinction between psychic-physical interplay and Spirit infusion. Bear in mind that Judith von Halle's variation on this no-food theme took place some thirteen years later than the Australian procedure's initiation, and simply happened to her with no intent of her own. And although the overall nature of the field-effect taking root and borne out in both experiences, along with the various descriptive names and resulting intellectual systems centering around this nucleus, are well within the boundaries of the general "Laski effect" previously described, a creative strange-loop effect is continually being demonstrated in each of these diverse forms.

While von Halle's experience was almost surely influenced by the related studies, hypotheses, and theories of Rudolf Steiner's long-pondered system—including what he termed the Mystery of Golgotha—a link with that very Golgotha event itself is apparent. Through resonant attraction, an actual temporal shift of physical-spatial sensing took place within Judith, for which phenomenon we have similar reports from other activities involving temporal-spatial displacements. In von Halle's case, however, what began as resonant attraction eventually became the full infusion of Spirit.

The Field Resonance of Golgotha

In that same historical period of Jesus and Golgotha, public crucifixions as a means used by the Roman authorities to terrorize into submission a near-suicidal Jewish rebellion, had increased in number until, according to one perhaps exaggerated account, crucifixions lined both sides of the main thoroughfare to Jerusalem for miles.

No matter how exaggerated, an overall reign of terror loosed on those hundreds of victims could indeed have been seared into the whole historical psyche of the human and added to the power of the Jesus event in our species' memory. (To get an idea of how such a field-effect forms, is kept alive, and can grow in power, go to the Holocaust Memorial in Boston and simply sit within close radius of it. I sat down near that memorial unaware of exactly what those great glass boxes were, and found myself slowly enveloped by an unaccountable level of grief that grew until it led to embarrassing tears. I felt that I wept for the whole history of mankind.) Consider how, in the field-effect brought into play in von Halle, such an enduring practice, with its own powerful feedback, can be experienced in endless ways.

A similar effect may arise from the Christian sacrament of the Eucharist, the symbolic daily taking of the body and blood of that crucified figure into oneself. Repeated century after century for near two thousand years now, by untold numbers of believers, this observance in itself could well become a field-effect of great significance. Recall Darwin's claim that any action repeated often and long enough will become habit, and potentially locked in genetically. Revitalized daily by untold numbers taking that sacrament (the word sacrament, akin to sacrifice, means *making whole*), the field involved in the event could thus, theoretically, be revitalized by each communicant unawares.

Stripped of its distortions, Jesus's crucifixion-resurrection would also be a demonstration of a mind-self-personality's ability to survive death, a full social awareness of which could bring to question humankind's great stumble over its fear of death. Jesus's example does not stand for the abolishment of physical death, which would upset the very ontological constructs of universal life with its vast capacities and wonders, but rather the abolishing of *fear* of death through recognizing the mind as an emergent process from the body, and not subject to the same fate. Thus, the distinction between the body and mind-emergent can dispel the fear of loss of body—fear being that most crippling of all constraints.

A Cautionary Note

The "akashic field" is reportedly made of a conglomerate of various other fields, as we have been surveying here. But the tendency of some people to treat such a storehouse as automatically a source of absolute truth is questionable. Almost any product of "channeling" is treated as truth because it comes from the numinous, inexplicable source it does. We should remember that any event from this field-of-all-fields must necessarily be translated, through those hundreds of billions of neurons in our brain-body, into our perceived awareness, then interpreted by each conceptual system. So this translation is always unique to and subject to serious colorations by the individual bringing the whole play into being. Memory, with its often vivid imagery, is never a one-for-one playback, but generally a haphazard or casual composite.

With this potential for self-deception and distortion in mind, we must nonetheless consider a further phenomenon that from one perspective is a "distortion," but from another perspective is both creative and generative.

As a child (the last of eight siblings) I heard family stories covering the recollections of different family members concerning the two major floods we experienced in the small mountain town of Pineville, Kentucky, where I was born. I was two years old during the second flood, and my visual memories of that flood were (and are) vivid, since they concern scenes I had myself witnessed. Yet several scenes—every bit as vivid in my memory—proved to have occurred in the first flood, before I was born. Many of those vivid images from childhood were thus, clearly, pictures formed *in my mind's eye* as family stories were related concerning events before I was born. Yet I recall them as my own, as they are indeed in *my* memory, which is all I have to go on.

Resonance between events may thus be an even stronger element than "facts." Carol Gilligan, in her studies of early teenage girls, points out that "grandmother tales" have great value in giving a young person a sense of identity, continuity, belonging, and security. By whom, and where, and when could the factual reliability of those tales be established, and of what use would such establishment be? That such largely creative inventions could never be substantiated in no way diminishes the value of "grandmother tales" and the wide variety found therein.

There are surely collective memories (as Carl Jung might describe them) subject to the same haphazard assemblies, but which have no more guaranteed validity than my childhood ones. The branches from an ancient root may breed a constant variety of romantic-historical-mythical overlaid shoots, all of which would appear to their beholder as quite factual and substantial, since not of our making, but given as a grace.*

My own teacher in India, for instance, following his tradition, accepted the power and sanctity of the Jesus story and all it stood for, particularly as symbolized by the crucifixion, but he emphatically denied that such a supreme yogi (as Jesus would have had to be according to yogic tradition) would suffer physical pain from his sacrifice. Yogis throughout history, he pointed out, have been able to anesthetize their body, partially or totally, and for long periods, if the need arose.

We might recall the fire-walkers of Sri Lanka, and in India I have witnessed some bizarre forms of yogic indifference to what would have ordinarily been severe pain and/or major physical damage (which damage, at least, never materialized). And we need to remember that the word "suffering" originally meant allowing, letting, being willing. It's a bit of contortion to account for terms such as "suffer little children," or even "women's suffrage," except through the original and wider application of suffering as willingness to allow or let. From this standpoint the notion of Jesus' "suffering" to "vicariously bear our sins" and take on the mechanically reflective punishment we so obviously deserve, is seen for the silly, distorted, and rather cruel deception it involves. As though only by outdoing us in pain and anguish could he vicariously alleviate or absolve us of our stupidity and hardness of heart—this lies beyond even fairy-tale fantasy.

On the other hand, a real gem lies in that label hung on Jesus as the "suffering servant," one who serves us by allowing—allowing the Spirit to move through him as we have need of him to be. This greatest of service, which involves simply allowing or letting, is not only one in

---

*Consider how that singular Golgotha event drew a different interpretation from Indian yogic masters. Many of these sages have long accepted the spiritual magnitude of Jesus and his crucifixion as a major symbolic act of history. Yet they equally attribute to Jesus those yogic powers they insist he must have possessed, powers that would have or could have brought about quite a shift in the technicalities of the crucifixion.

which any of us can partake, but is the most active and powerful move through which we can open to and devote our life to our own and the planet's common good.

Distinguishing Between Mind-Fields and Spirit

In my way of thinking about Spirit, spiritual fields are markedly different from religious fields, and even my use of the term "fields" to get into this subject of Spirit is misleading. There is, I wager, no such phenomenon as spiritual fields. *There is only Spirit.* Spirit is not a field, nor field-effect, nor does it manifest through fields or field-effects. Meditation may offer a possibility of opening to Spirit, but is generally undertaken by a person with a personal, if hidden, agenda; such an agenda inevitably influences the nature of the meditation and subtly blocks Spirit.

While there is no spiritual field, and Spirit is not subject to or part of any field (akashic or otherwise), Spirit does manifest in a myriad of constantly changing, shifting ways, according to the infinite phenomena through which, or as which, Spirit manifests. Spirit is not perceived as such, but is a resonance we must attend only as a possibility to be allowed—should it manifest. Only Spirit within us can perceive Spirit as "outside us." Spirit is not reciprocal in any way. Spirit has no locus; it is universal, both within and without. We do not "enter into Spirit" in any fashion.

Spirit is not strengthened by our participation, reception, or response, nor is it in any way dependent on us. No religion or belief system can be built around Spirit. I cannot "practice spirit" until it becomes habitual. Spirit never stabilizes into a ready reference point, even though our constant and continual experience of it, should such occur. Spirit is the only phenomenon in our life that is truly a "river we can't step into twice," since the movement of Spirit is just that, a movement wherein truly "no man knows its comings and goings—it bloweth as it listeth."

Further, Spirit may be the only toehold, or mind-hold, that the mind has for grasping—or at least getting some inkling of—that unknown we term the *Vastness,* into which Spirit seems to merge and from which Spirit seems to flow, a glimpse of which may come if we allow an openness to Spirit sufficient to sense its general course and direction. Even so, this results from our openness and our sensing, not Spirit's. All of these observations, of course, are my own conjectures, certainly not Spirit's, which may or may not have much to do with that to which the word *Spirit* refers.

Cultural Counterfeits

Notions of a "marriage between science and religion," or worse, science and Spirit, thus making the religious or Spiritual response a "scientific" fact, is quite popular today (and rather a form of intellectual incest, since science is itself a most powerful form of the religion we invest in and invoke). But any scientific verification of Spirit is a hoax, although perhaps a comfortable one, of which there are legion. Spirit is neither an energy nor power available to analysis or instrumental detection. Our wise man of two millennia back observed that God was a Spirit, but he didn't say Spirit was God. The first is a relationship embracing us all, the second is an equating with—which may be stretching the boundaries. That is, he clearly implied that we were of that Spirit and God, a product of—not necessarily identical to—its "process."

Spirit has perhaps one manifestation (a word that isn't necessarily appropriate, but the best I can do) that might be resonant with any of our sensory patterns or mind-sets, and so can be considered as perceived, after a fashion. That is Silence, as introduced by Robert Sardello. In this Silence there is nothing to be perceived, at least not in any ordinary way (here I capitalize Silence to distinguish it from simply an absence of noise). I would further point out that Silence is an intense involvement—perceived, one might say, as a precarious and delicate balance of mind—the intensity involved being, perhaps, a form of "soft will."*

In Zen archery, for instance, the master is "breathed by It," the Spirit, and at that moment the (ninety-pound) master lightly holds the bow with limp muscles, and the powerful bow—which only the strongest of men can bend on their own—bends without the master's muscles being involved at all, other than going through the appropriate motions. "It" bends the bow, perhaps as an extension of the master's will, which seems almost an independent phenomenon.

The power within the hara can go either way, physical or heart-aligned, and there is no higher arbiter determining which direction is best, since "best" is an intellectual evaluation-judgment, not part of the hara equation (nor of Silence, or Spirit, which are of a nonphysical state falling outside both Hara and any terminology we come up with).

Opening to Spirit, Opening to Silence

George Fox, whose life brought about the Quaker Movement, opened to Spirit without a personal agenda—the only way in which Spirit can be opened to, or even approached. This openness indicates a "soft will" surrendered to the heart. In Fox's case, through his soft will he opened to the heart while embracing a "hard will" able to follow the heart regardless of where that heart and its soft will led. A rare combination—George Foxes have not been overly plentiful in our history.

Robert Sardello describes his experience of Silence as it manifests in Spirit, or perhaps even as Spirit. Sardello's account of Silence is a description of *his experience of Silence,* and never construed as a description of Silence itself (as though "it" might break "its" silence to tell us

---

*Soft will is the capacity of the *kath, chi,* or *hara* (that ball or bundle of will and power right below the navel) when surrendered to the heart. Purity of heart is to will one thing, our existentialist Kierkegaard pointed out. Some Eastern disciplines concentrate on a "hard" will of power and strength manifesting on a material-physical level, which is equally possible and equally legitimate.

about itself!). This rules out conjecture and hypothesis and keeps us grounded in the experiential phenomenon itself, as opposed to intellectual, mind-wandering theories or romantic longing. As Bernadette Roberts insisted, we have only descriptions of *our experience* of God—which cannot be construed in any way as "God's experience," nor in any way as a definition of God.

This is not nit-picking semantics, and the same holds for Silence as for Spirit, which are almost surely the same. Sardello's is a description of his own experience of Silence, which is of great value since it is a tangible and living guide we can follow. Sardello's account lends, perhaps, to a phenomenology of Spirit, which one can really grasp only by having experienced that phenomenon in one's own self. Sardello's living account as an example gives substance to that phenomenon, making it available to us if we open to it and allow. Sardello's remarkable little volume titled *Silence* subtly points the way by which we too might become aware of Silence, and so, perhaps, be open to Spirit. Even so, we don't "invoke" Silence as a shaman invoking a spirit within his tradition, since this indicates such spirit is available through such invocation and tradition, throwing the issue back into the religious category. There can, however, be no case in which Silence is *not* present. Our awareness of that presence is another matter, not available as a casual pursuit or pastime.

Enter the Problem of God

Our discussion of matters seen and unseen, like the issue of self, touches on cosmology, ontology, and creation, and sooner or later tangles with the concept of God—an ancient abstraction of mind, which indicates, at best, a verb (not a noun), particularly to the extent God is considered as, or involved in, creation. Here we come to the mirroring of all mirroring, the "granddaddy" of all strange loops: *Creator-and-that-created give rise to each other.* Creation does not "exist" as some superior form of action or actor. "To exist" means "to be set apart from," and this fluid verb-action of creation cannot be set apart from itself. *There is nowhere to go.* If there were such a "where" to go to, all we would find would be the same process—"itself" all over again.

We can then use the verb *creation,* indicating action or movement. Creation is a "fixed" name we can agree on and refer to, such as *God,* but in this case the naming does not freeze the action or movement itself into a thing-object-entity molded by our nomenclature. And this is important: the name for some phenomenon enters into the overall neural fields organizing that phenomenon into a perceived event, which is why "speak the word and it appears" is true, at least in our mind's eye of imagination.

While this book hasn't overly dwelt on the "problem of God," the issue hangs on the horizon at all times—by default. So to such a problematic issue we will turn, not with the illusion of solving the problem, but of opening to it without expectation and allowing what might then take place.

ELEVEN

·■·

THE PROBLEM OF GOD

"Fire! Fire! Not the God of the philosophers, but the God of Abraham, of Isaac, and of Jacob"

(So scrawled Blaise Pascal at the height of his great mystical experience, one bringing metanoia, that fundamental transformation of mind and heart that changes everything.)

In a splendid essay on "The God of Abraham," Christopher Bamford notes that the name for Abraham's God, El Shaddai, translates as "the many-breasted one," and true to its name is feminine in both its grammatical use and general syntax. That the God of this "Father Abraham" of Hebrew history—from which arose Christianity, two Testaments, and other romances, bringing endless generations of argument and warfare—was originally a female, and therefore a "goddess," is surely noteworthy, though generally unknown to Christians.

Creation itself is, however, neither a subject of argument nor grounds for constant warfare, so one invents a male god for that. Which is to say the God of Abraham, Isaac, and Jacob, whose sacred fire forever changed our beloved French philosopher Pascal, eventually was replaced with the fire and brimstone figure of Jehovah—plenty of male dominance and warfare there, and a major example of the cultural counterfeits shaping our history and bringing grief.

An Early Cultural Counterfeit

Note that Abraham's female god with *many* breasts was obviously able to suckle and nurture an endless number of tribes simultaneously and indiscriminately, as mothers tend to do with their offspring. All were equally El Shaddai's children. Creator *and* Mother of Earth, She is our model and fountainhead of creation. She has long been presented in the West, however, through this strange masquerade of a father-creator, one who could not suckle and would hardly have had time to nurture, since roaring into history to whip into line and make that Mother's offspring toe that line—which is more in keeping with a male-dominator culture, to say the least. How we lost Abraham's nurturing God(dess) and were left with only this demonic Jehovah and our lament of "feeling like a motherless child," is an interesting and speculative story, but far too complex to tackle here.

Equally intriguing is a link between the Gods of Abraham and Jesus, arising in references Jesus reportedly made to Abraham, in the native language Jesus spoke. His language would have been, according to scholars, Aramaic, not Hebraic, and surely not Greek (wherein one Paul-the-Apostle makes his Greek-oriented Jehovah top dog in his virtual-religious world, to which Jesus was unfortunately linked by our cultural history).

Anthropologist Mircea Eliade observed that great myths form only around an initial kernel of truly great people—giants of history, not mickey-mouse or limp milquetoast characters. And indeed we find a true giant of history buried beneath those two thousand years of mythological overlay on Jesus, sadly centering around, relating to, and encompassed by the worst aspects of this fire-breathing mythical Jehovah.

In my life I have had any number of vivid and transformative experiences resonant and associated with this romanticized and mythically overlaid Jesus, and I find myself creating romantic overlays of my own, adding to those this magnetic and magnificent figure has attracted down through the ages. This compulsion on our part to romantically-mythically overlay such great figures, generation by generation, shows the power of both mythical overlay and of such great beings.

My own longings, aims, and spiritual ambitions found their greatest impetus and organizing nucleus in that mythical-historical figure. And so for me the reality of Jesus lies within my mind, heart, and personal history with him, and I couldn't care less about the dead annals of a supposed physical, calendar-dated history culturally agreed on. Nor about a dull theology, which has not yet displayed to me anything near what I have known personally, and felt brimming forth from this incredible historic-mythical figure.

To add a bit to this Jesus overlay, note again that the language he used was apparently Aramaic, not Hebraic. The Aramaic name which Jesus used when referring to creator, creation, life, and growth, was like Abraham's *feminine*—and El Shaddai means not only the many-breasted one, but that one on whom we can "lay our head and be restored"— hardly so masculine a title as "Father," as bestowed by Greek-influenced latecomers on the Jesus scene, such as Paul: this, of course, being the same Paul on whose radically overlaid and distorting Greek version the whole of Christianity rose to power on false wings.

Consider how such a masculine title—*father*—for the function of creation may have been quite inappropriate to these earlier Aramaic-speaking people, creation and giving birth generally being female functions. To a pragmatic and practical Hebrew, the idea of males giving birth, regardless of to what, to whom, where, or when, was surely an oxymoron, bringing serious cognitive dissonance. Since arguing over the fine points of scripture was (and is still, to some extent) virtually the lifeblood and sinews of the brain for the Jewish culture, scripture was like a whetstone with which they kept their wit and wisdom sharp. As my friend David Tetrault pointed out to me, Hebrew scripture wasn't necessarily a code of conduct or verbatim history, so much as a way of presenting our ever-present paradox and problem of self to ourselves—a problem over which Jews could argue, as they have for centuries. We might to some extent emulate this tendency, to our advantage.

According to traditional, established and accepted "sayings of Jesus" (published in that final writing of a "New Testament" long after his death), this greatest of great beings is reported to have proposed that "the time has come and now is when one does not worship God on a mountaintop or in a temple, but in spirit and truth." This rather eliminates churches and political alignments, taxes, armies, and prisons (indicating that this quotation was probably accurate), as well as replacing something substantive, as mountaintop and temple, with a most perplexing abstraction demanding definition and explanation—and a bit of the heart's intelligence—to be grasped (giving even more authenticity to the reported statement).

This *God-as-Spirit* has not only a feminine designation, but involves a Darwinian biology through and through. This living body of ours is the only temple wherein that truth can be found, which offers us, however, the key to a genuine cosmology-ontology of creation, if we keep it free from religious-political overlays.

Declaring that God is a Spirit, Jesus referred to this Spirit as a "whole," a phenomenon that cannot be broken up, or set apart from itself, and further, that it moves as it pleases. Our reifications are idol-making imaginations, where our fancy thinks to influence the phenomenon through our endless litanies of babbling prayer and plea-bargains. These noises simply aren't heard. There is no one "there" to hear us, since our babbling reflects our incoherent thought-forms spoken, and the Spirit addressed is pure coherence.

Suzanne Segal, in her little book, *Collision with the Infinite,* gave us a stunning, mind-stopping clue and truth about self, Spirit, and God in her simple statement: "The Vastness doesn't know anything is wrong." The Vastness to which she refers is the state of pure coherence, which is resonant with Robert Sardello's Silence, a state we can become aware of, somewhat, only in our own state of pure coherence—as best we can manage such a near-impossible achievement. All of which is to equate Vastness, Silence, and God as the process-phenomenon behind everything. This process-phenomenon is simply not available to anyone as an object for analyzing or experiencing as a *thing,* nor is "it" comprehensible to a mind with any trace of incoherence lingering around. All one can do is to experience what such total coherence means by achieving it. Even then one cannot turn and analyze, converse about, or be objective about it, as though it were an object for study.

In regard to our current best-selling titles about prayer, William Blake once noted that "as the plow follows words (the verbal outpouring of the plowman) so God answers prayer"—which is to say, we think we can talk oil and water into happily blending, but language has its limits. Prayer, whether answered by God or random chance, became for a while a popular New Age conference subject, projecting onto Cloud Nine the theatricals of a truly critical issue—the field-effect that group thought, if fairly uniform and coherent, can have when targeted on a single object or subject. (One version of this is found in the odd practice of Tibetans who chant over the body of a dying brother, a subject touched on in this book's closing.)

Abraham's feminine definition of this creative force is an argument strengthened by Ashley Montagu's superb biological study, *The Natural Superiority of Women.* In Montagu's study can be found, perhaps, reason enough for the timeworn jealousy men have for women, and for their continual attempts at takeover of all such natural female processes as intuition, ancient wisdom, "thinking for the left hand," knowledge of the heart, intelligence for birthing and nurturing offspring, nurturing each other, and having compassion on ourselves. Those male-oriented power-plays to wrest such quiet power from Mother Earth, creation, and women in general, have upset the natural order of things, bringing suffering and chaos in their wake. This disaster is constantly increased by this same male compulsion to gain *control over the very chaos the male intellect brings about,* making matters worse and worse—as found in Goethe's *Apprentice* in absence of his Sorcerer-Master. Thus arises our ever-present dominator male culture, ever trying to re-establish order through their harsh legalisms cut off from the heart's intelligence, and always bringing but more chaos.

Abraham's Movement

Examining the relation between Abraham and his God(dess) we find Abraham's "blind" obedience to that Spirit (as I assume it to be). The Spirit says, "Move," and Abraham *moves.* He doesn't ask to where or toward what he should move, he first just moves. Out of that action of movement comes the means *for* the movement itself, the directions of where, why, and when to move, and the substance of what is found therein. This puts the issue of Abraham's obedience in the strange-loop category.

The movement of obedience to Spirit *creates the where and why of that movement,* the same response found millennia later in Meister Eckhart, who "lived in wandering joy, without a why." To live without a why is the challenge of all challenges, and the root of forgiveness as a primary force in line with creation. Eckhart also claimed that "without me God is not." And of course he observed the opposite as well, all these being expressions of the mirroring strange loop phenomenon and variations of Abraham's example.

In my first book, *Crack in the Cosmic Egg,* I commented that one cannot step out into nothing, since there is no such thing as a no-thing. Stepping out into apparent nothingness, *one finds something always forming underfoot,* for our awareness can bring about the creation of something to be aware of. This formation forms only as we step out into it, however—a step which, at its moment, seems the equivalent of jumping off the cliff without a parachute. But the stepping out—and only that stepping—creates what forms under our step. That is, true "leaps of faith" can be creative acts.

A prime example of Spirit and its movement is found in George Fox, from whom the Quaker movement arose. Were Fox at any time or under any circumstance, moment by moment, faced with a decision (the word means to cut off alternatives), he simply stopped and waited. At some point Spirit would move him as he needed to be moved, bypassing thinking in order to just move without hindrance. All too often Fox's obedience to Spirit led to prison, where he spent a considerable part of his adult life, but that, too, is illuminating. Culture has no use for the nonpredictable whimsies of Spirit. Temple and mountaintop are far safer cultural alliances, and a dungeon is a far safer place in which to keep the George Foxes of history.

There is another aspect of Fox's willingness to be moved, even to prison. One might ask if this willingness doesn't contradict the definition of the intelligence of the heart as that which moves for our well-being. Well-being for one in *service* of the heart is a "universal" issue, as heart is not personal. One's personal self has given over to that universal self, which is the heart's nature. Intellect in the head, whose "well-being" without reference to the heart's intelligence, generally proves disastrous in the long run. Fox's suspension of intellect and personal investment on behalf of Spirit's wholeness is an integrated movement, its wholeness not necessarily in keeping with intellect's idea of well-being. So Fox's well-being rested in Spirit's wholeness, regardless of where it led, placing Fox and prison in the same position as Jesus and Crucifixion, the extremities between the two being beside the point here. Both Jesus and Fox, as history would show, have led toward the well-being of the overall history of us humans, whether or not we each open to such in our own life.

Discoveries from the Left Hand

My reading of Fox's *Journals* led me to the discovery of Jerome Bruner, to my mind one of the twentieth century's richest thinkers. One day back in the early '60s, mired in the endless drafts of *Crack in the Cosmic Egg*, my romanticized inner model-image of George Fox stirred me to remove to the college library where I determined to discover what Fox practiced (and perhaps break my temporary stalemate and slight slough-of-despond). I found a spot where I could simply stand and wait, in good Miltonian fashion, as I assumed Fox did, without appearing unduly strange.

Doing what I could to suspend my ordinary roof-brain chatter, I waited, silently, for "it" to breathe me. And in what may well have involved some aspect of the Laski effect, "it" did indeed finally move me. What I can only call a force moved me quite physically, as though I were drawn like a puppet on strings to a section of the library I had not visited before (and I was as relaxed and willing to be so moved as I assume a puppet is, a rare state for me). I went straight to a particular shelf somewhat above my head, where I reached up and took a book, which seemed to me at the time to literally fall out of the shelf into my hand. The book was none other than Jerome Bruner's *On Knowing: Essays for the Left Hand,* which moved me out of my stalemate and gave me a better image of what I presumed to be Fox's practice and some of the rewards therein. Further, Bruner gave many an insight into our human Spirit and its "left-hand operations"—those forces which function beneath our usual "right-hand thinking" with its "left-brained logic."

Through such forces, giving rise to creation and birthing, we are not limited to a choice between reification and abstraction. We are dealing with aspects of the strange-loop phenomenon, which is neither defined by nor confined to our mental constructions. In such looping lies a way out of our all-too-human dilemma—today as always. At stake in this movement, generated outside our ordinary mental process, is a "single centimeter of chance" given us in which to respond, as Carlos Castaneda's don Juan points out. The instant's opening of Spirit is subtle, swift, and easily overlooked—or not perceived at all. But at that instant's opening, if we immediately and instantly give over to it without the slightest flicker of hesitation, that force can move in and become the principal issue of that event.* In opening to Spirit one isn't presented with a fact or possibility for pondering and deciding over. Our response must be as instant as the possibility

---

*This is an example of "unconflicted behavior," as referred to in various parts of this book, originally documented in *The Biology of Transcendence.*

that is opening, making them virtually one action, not two, and may require a bit of practiced attention. (Again, as Blake said, "Mechanical excellence is the vehicle of genius.")

The first impulse, in the Bruner book incident of mine, was so subtle as to be easily dismissed, or simply missed altogether. Only due to the fact that I had waited, rather suspended, for something of which I really didn't know what it could be, could I have ever felt so subtle a movement. Nonetheless such subtlety was instantly powerful, so long as it was not doubted.

Later in the 1960s, winding up my writing of *Crack,* I was in the library late one evening checking references I had made. On leaving and walking toward the exit, I felt a sudden pull to my right, into a dimly lit room (apparently closed for the night), directly to a table whereon was a row of new books waiting to be cataloged by the librarian. The pull was quite extraordinary by then, and I compulsively reached for a green, hard-bound volume. Picking it up to read its title, a cold chill ran from the base of my spine to the crown of my skull, my hair (which I still had back then) literally standing on end. I clearly perceived being in the presence of the uncanny as I read the title: *The Teachings of Don Juan: A Yaqui Way of Knowledge,* by Carlos Castaneda. This was a new and unknown work from the University of California Press, which subsequently brought a major shift in my book, *Crack,* and insertion of a new chapter, "don Juan and Jesus."

There is an admonition in the sayings of Jesus to keep awake and alert, for we can never know at what moment "it" might come. This coming of "it" covers a wide range of possibilities, and is ontological and universal rather than limited to any religious or philosophical notion or event. The issue is just what such alertness means. The preparation for such alertness seems to require being in that open state without reserve, an empty expectation, which alone allows for the moving force of Spirit, rather than that avalanche of "what-ifs" and protests our logic will immediately impose.

But in all practicality, one asks how can we remain in such openness and tend the world-out-there (echoing Simone Weil's fundamental issue of grace and gravity)? The answer lies in "not letting the left hand know what the right hand is doing," as our great Model pointed out. This is no simple matter, but rests finally on which "hand" we hold to be of primary, supreme importance—world of folly or world of spirit. We have to be attentive on two different levels of consciousness, temple and mountaintop on the one hand, Spirit and truth on the other. This can throw the issue into a strange loop wherein we can serve and live as both left- and right-hand worlds give rise to each other.

As with my being led to discover Jerome Bruner, I was led to this discovery of Castaneda by a resonance brought on by my absorption in my own book that whole evening, and my otherwise "blank state of mind" concerning such routine matters as winding up for the night to leave—a "right-hand" affair leaving my "left-hand state" open.

In summary of this, if the Spirit says move, and we move there and then, the movement of Spirit in our life "makes all things new moment by moment." We then live in a state of "constant astonishment," as my friend David Tetrault speaks of it. The least bit of resistance and the presence of Spirit simply *isn't*—no one the wiser, not even oneself. On a moment's reflection here, we can see a direct correlation between this open alertness and the blank mind state involved in all Eureka events, which is resonant with that capacity-compulsion of the very young child for open-eyed blank staring—which should be a capacity never lost. So long as mind and its busy intellect take up the stage, creative discovery cannot unfold, and the same holds for Spirit.

The answer to our enigma is to open to and live in a dialogue between the two, left and right hands. Irina Tweedie referred to this stunt as similar to "walking a hairline thread over a chasm of fire." (As for me, I lose my balance, and the older I get, the worse my balance. I stagger.)

Bear in mind poet Blake's statement: "Mechanical excellence is the vehicle for genius." The instantaneousness of all such appearances and creative moves rests at some point on this mechanical excellence, which we must have developed, yet which ironically must be left behind the instant genius manifests: Kekulé's snake, Hamilton's quaternion flash, Mozart's round volume of sound—all are like Blake's "eternity in an hour" revelation. Revelations do not hang around that we might decide to heed them. Mind must have done its homework, developing the "mechanical excellence" that is the servant of genius. Realizing why the mechanics cannot be carried over into the appearance of that genius— or there is simply no genius—is not complex. Genius can be blocked by the very mechanism it must have for its appearance.

Again, a resonant dialogue between body and Spirit, grace and gravity, is called for. The trick is to know when to work, and when to do nothing except open. Robert Sardello's suggestions for opening to Silence and nurturing such an opening through a kind of radical "not-doing," is a neat summary of this living in balance. There are other cues. We are hardly without helpful hints, once we are ready to allow and follow this subtle, intuitive heart-intelligence—heart being the gateway to all things new.

TWELVE

"ABANDON HOPE, ALL YE WHO ENTER HERE"

Walker, your footsteps
are the road, and nothing more.

Walker, there is no road,
the road is made by walking.

Walker, you make the road,
and turning to look behind
you see the path you never
again will step upon.

Walker, there is no road,
only foam trails on the sea.
ANTONIO MACHADO

This book has described some of the many ways in which humankind has cut itself off from its original matrix of the heart, and thus no longer serves as the heart's new way of thinking any more than it serves as nurturing caretaker of Sophia, Earth's Spirit—both of which we are designed to do. We are equally involved, along with Sophia, in the reality states brought about by such a threefold interaction of the mind, heart, and

Spirit, on which both we and Sophia depend. The results are in turn dependent on us, in a mirroring loop of intricacy and balance, though trying to spell out such effects may at times sound like fantasy.

I have also suggested that fields of potential can be intelligent, creative, non–temporal-physical, and apparently permanent in whatever formations those fields bring about, or are brought about by. Such fields are then available to us so long as we interact and participate with them, personally and at times universally. Recall the round volume of sound experienced by Mozart, "called up" in effect, and replayed by my university pianist-professor two centuries later. Such "calling up" may be the mode for a viable, functional matrix-of-mind as needed by minds like ours when, on losing our physical matrix and plunged willy-nilly into a nonphysical realm, we might find ourselves in a matrix of like "called up" order.

To speak of evolution's "work-in-progress," then, is to point out that such work is ongoing, very much present with us, available, with conscious study and attention, to our entering into—even now, in our ordinary makeup. That is, from this somewhat latent capacity we can establish the foundation or general outline of a non–temporal-spatial yet stable "next-matrix."

That some semblance of such a future state must be established in our current one is rather like the strange loops in brain development, as discussed in the early chapters of this book. Even more potent, perhaps, is the resemblance to an evolutionary blueprint of possibility, as in a gene, which can act as a guide or outline for a creation of our own, although no content can be genetically included, and we must seek out or create our own. This may very well be the way our next level will open to our content search, since such forms may always be opening and filling with such content.

Transcending Hope

One might think that through this evolutionary potential to override our limitations, hope should "spring eternal in the human breast," as long quoted by romantic optimists. Yet my late departed friend, George Jaidar, spoke of hope as our great enemy, our nemesis and downfall. Jaidar's observation was not calculated to win friends and influence people, nor pack the halls with New Age enthusiasts. Didn't Dante's sign at the entrance to hell read, "Abandon hope, all ye who enter here"? To make such a cheerful title here to welcome a reader to this closing chapter is hardly calculated to please a publisher either.

I took years discovering what Jaidar meant—that giving up all hope is the way beyond despair, the dead end to which hope always leads. Hope and despair are a negative strange loop, each giving rise to the other in a deadlock. Hope cloaks and hides a hidden agenda within our mind, an ace up the sleeve we secretly carry that qualifies the openness Spirit must have to move into our present. We are then left with some variation of our past, rather than that which makes all things new, as our Great Being described. Our life preservers sink us.

Alert to such false hope, then, we turn yet again to mind—fragile to establish initially, but subsequently more powerful than the simpler awareness of lower species, from which mind-as-self-awareness initially arose. This self-aware mind is itself fragile in comparison with a higher form of being—emerging in us now—which can go beyond any limitations and constraints at that new level. *This is evolution's work in progress: a matrix for mind when the body goes.* Were we humans to have hung on to our mammalian system in fear of losing it, we would have lost the only way to move beyond it. And the same holds for movements beyond this current phase of ours, in which many might be called and few chosen, since the necessary selectivity is still at work.

Heart Instructions

The ramifications of Steiner's speaking of "the heart teaching mind a new way to think," will give some understanding of the next level of this work-in-progress. This is "work" with which we must be involved, since we are an integral if not major part of it, as well as recipient of it.

Bear in mind that this matrix-goal is always "in progress." Our wisest of sages observed that "the Son of Man has no place to lay his head," as have the animals. In giving up a "place to lay his head" he moves beyond the need for such a place. This is the gist of evolution's work in progress within us, always preparing to move us beyond. Were such progress finalized, giving us a "place to lay our head," there would then be no evolution or creation, only stasis and stagnation. In true creation, only the movement of relationship exists.

So work-in-progress-toward is what our life is about. This hardly proves to be "work" in our victimized sense, but a joyful discovery, as observed in toddlers, if allowed. Constant astonishment always leads beyond itself, so "toward-which" the design can go is never completed, but open-ended. We are always toddlers on some level, by which our universe expands in its ever-moving-toward.

Bear in mind that *matrix* means "source" and, as with our word "mother," implies nurturing or caring for. In our case some universal or higher nature of heart is, by default, our nature as well, and we recall Darwin's claim that such nurturing, expressed as love and altruism, gave rise to us, lifting us out of our animal heritage: or is trying to.

In pursuing Steiner's proposal of heart teaching us a new way to think, we find that our current thinking, such as it is, can (and must) be an active and integral part of that learning. Since we are equally heart with intelligence, as well as brain-mind with intellect, we can only begin in this place and at this time—Now—to open to a conscious awareness of the process leading beyond.

Illusions of Paradise

Our religious reifications of what "should" happen after the death of body-brain might imagine a stable, paradisiacal state, which, could such anomaly take place, would be as static and immobile a state, fixed and frozen into itself, as Dante's seventh circle. What is found, however, in our pursuing this next matrix for the mind is *that the pursuit is all there is.* Like Abraham's movement, following the dictates of Spirit to move—within that movement the whole issue lies. In the movement is the ever-present answer.

So that next level of the mind's matrix is not only the matrix that the mind must have on leaving this current physical one; we must be actively involved in the unfolding of this new thinking itself. Such involvement-action is not only critical to establishing such thinking, it is part and parcel of that next move and our ability to take part in it. It is like Roethke's "walking by falling forward," with its corresponding question: Where is there to go? The answer emerging is: "Keep going."

As Shakespeare said, "If you would possess a virtue, you must first assume it." Acting as though you possess some capacity plants the seed of that virtue in us, if only in imagination. Acting out our pretense can bring about the ability-soil in which the seed can grow. The model, imperative to any new growth, is formed by our action, as is the growth that follows.

All of which is to say we must actively enter into the creation-formation of the matrix for the mind needed when body-brain-heart must be gone beyond. Even from our first attempts at such an opening of heart's new-mind, that state or capacity plays a role in maintaining our mind's integrity, itself needed in such movement-pursuit. This integrity (or coherent wholeness) must be firmly rooted in our *current* situation—our physical matrix—if such integrity of mind is to later hold in the often disorienting and fragmenting transition from our current matrix into a nebulous, intangible, and unknown one. Among other things, this surely means we must work toward restoring the coherent governance of fore-brain and heart, and cease to live according to the reflexive whims of the reptilian hind-brain—which, like it or not, is in fact the way most of us currently live.

Instructions from the Non-Ordinary

Herein may lie the greatest value of the paranormal or non-ordinary episodes that have filled history. They are like Eureka forecasts awaiting an actual breakthrough into realization. There are several ways in which paranormal phenomena might be engaged in creating a matrix for the mind following the death of body. Consider the numerous examples given in just this book—Kekulé, Mozart, and von Halle among others—wherein the mind's reach is far beyond current or past scientific validation or public acceptance. Any of these effects occasionally breaking through to us, even in the great welter of noise and incoherence at every hand, would almost surely always be available to a mind already lifted outside this maelstrom of our overactive ordinary reality. Mind separated from the brain-body could easily "tune in to" areas of resonance with which the mind had already established resonance in ordinary life.

Resonance is not limited to or necessarily dependent on electromagnetic fields as found in the body-brain. There are many ways in which we attract forces, resonances, or field-effects to us automatically. Many of the mind-disciplines found in ashram life, Sardello's exercises in becoming aware of Silence, walking a labyrinth and so on, can set up resonances—and the intent of such pursuit can be broadened. The Monroe Institute discipline leading to a phenomenon called the Park, detailed later in this chapter, is quite striking in this regard.

In my twenty-second and twenty-third years, I experienced brief periods of physical immunity to fire, pain, or harm—exhilarating and having no after-effect. Similar episodes had led to my discovery and practice of *unconflicted behavior,* with its suspension of ordinary cause-effect within a select and narrow range. I have touched on this possibility in several writings, most recently in *The Biology of Transcendence.*

Concretizing a Non–Temporal-Spatial "Illusion"

Further possibilities of nonordinary openings can be traced out in the documented account of mutual lucid dreaming practiced by two college students in San Francisco back in the 1970s. Although the couple was involved emotionally, they lived separately, at home with their respective parents. Becoming interested in lucid dreaming, they read up on the subject and, following standard procedures, were successful in the pursuit. They dutifully made copious notes on their adventures, sharing and discussing them with each other at length. (Such sharing tends to concretize nonordinary or dreaming experiences. Back in my academic days, six of us teaching in two adjacent colleges found, in conjunction with Jung's dream practices, how these experiences grow and expand the more attention paid to them.)

After a time, it occurred to the couple to try such lucid dreaming *at the same time* with the intent of meeting each other in that state. And indeed they worked out the procedures and found themselves together, equally present and freely conversing with the other, in the strange lucid dream settings that formed. In the private other-world they had experienced *separately,* such dream-settings, while vivid, were unstable and easily faded, as dream states ordinarily do. In their shared, mutual-dreaming states, however, not only did they perceive each other as they ordinarily are, they conversed, exclaimed over, and discussed what was happening at each moment. Everything in this *shared* dream world was stable, tangible, and totally present to all their senses. Their ordinary "wake-state-minds" seemed intact, even as they moved together into a series of purely imaginary realities.

Any particular locale they explored would continue to extend spatially, beyond and in front of them as in natural settings, no matter how far into such a world they moved. (It took awhile for them to relax into these states and not be apprehensive, as at first.) Their verbal and separately written individual reports of their shared experiences agreed point by point.

The most intriguing aspect of all was finding that the particular state-environments they explored together were permanent, and could be called up again at a later time and entered into, just by remembering and thinking of the event, whether brought up again by both of them together, or either one alone. Each time, this apparently stable yet dreamed state was found and entered into exactly as they had left it previously. Consensual agreement on or within a nonordinary state could give that state the same stability and apparent permanence as our ordinary reality.

Hold to this particular aspect of their adventure, and recall our discussion of field-effect and the odd fact that apparently any field aggregate brought into being can become a stable organization when shared with others, and take on permanency. The key given by the young couple for such possible entering and experiencing of shared fields should be obvious, and it lends to our knowledge of the open-ended possibility for matrices of the mind beyond physical worlds. Remember, so long as the young couple in the above example experienced their states alone, the states were neither stable nor permanent. Shared—which means equally created—the states were "there," wherever "there" might be. Thus, "when two or three are gathered together," a third force may be called into the situation to add a stabilizing factor. David Bohm intimates that in true dialogue between two people, such a third force is automatically present.

That on being shared, such states take on the permanence and stability of a consensus reality suggests that such a consensus between two or more perceivers on what is being perceived enters into the possible stability and sustaining of such a nonordinary state. So the proposal of bringing about or aiding in bringing about, a nonphysical "matrix formation," available in a non–physical-temporal state, is certainly feasible. The word matrix means "source," after all, and generally implies a supportive or nurturing state associated with our heart, where source can be fluid and nondeterminate.

In summary: Odds are strong that a nonphysical mind-field can be called up in a physical state, which is at a considerable remove from the nonphysical mind-field. In which case, why should the nonphysical mind-field not be called up far more easily by a mind in a nonphysical mind-state to begin with? Rudolf Steiner proposes that making a thorough account of our life gives us an ever-more stable frame of reference, a solid foundation from which to function in ordinary life, or the nonphysical domain, as the case may be. Robert Sardello suggests starting such a discipline by a nightly review of each day's events, recounted to oneself in as orderly a fashion as our memory allows—and such memory involved grows stronger with usage.

The preceding suggests that our ordinary daily reality may be maintained by consensual agreement as well. Such agreement, inherited and brought about in every new life, would continually strengthen the field-effect of consensus reality. Each infant-child and creature, in creating its own structures of knowledge as outlined previously, would automatically increase and stabilize those "aggregates" of phenomena involved, in an akashic field or otherwise, as Steiner and others have discussed. Needless to say, such stabilization would feed back directly into our ordinary experience of consensus reality, making it stronger as well.

Perhaps, as proposed in that strange and disturbing volume called *A Course in Miracles,* this field-effect and the strange loop of Creator-and-That-Created, brought in at infancy and etched into our neural patterns, is far more plastic than we are led to believe, and can be changed by a change in our attitudes and practice at any age.

Possible Signposts in Evolution's Unfolding

A number of experiential strands of nonordinary phenomena are similar or resonant enough with this proposed non–physical-temporal state to suggest, in substantial though hypothetical fashion, a general direction, which our creative evolution brings about, if we allow it.

First, on being separated from our physical origins, we are left as the *mind only.* The critical issue would lie in maintaining the *integrity* of that mind—its unity and coherence within itself—when all ordinary reciprocal checkpoints of orientation are not there. Consider the problems the mind faces, and the peculiar solutions it sometimes creates, on prolonged sensory isolation. In an earlier work I reported the experiences of two coal miners trapped in a small airspace deep in the Earth for days, and the quite transcendent, near-mystical states they eventually shared. John Lilly, in his flotation experiments, kicked into a more intense frame by ketamine, experienced a wild and varied array of such imaginary but tangible states. Jean Houston brought about astonishing experiences in her (drug-free) mind-altering devices two decades or so ago (which I dutifully, in my skepticism, underwent—and was astonished by the richness opening to me). These in turn bear similarity to controlled forms of lucid dreaming, which are of an even more intense reality since experienced in an awake, not sleep, state.

Indications are strong, however, that on leaving our shared physical matrix, the mind tends toward dissolution, as originally threatened people in long isolation in flotation tanks, for just such lack of contact with a stable point of reference. Claims have long been made that some minds, on separation from the body, find themselves literally "homeless" and wandering in confusion. A sister of mine gave every indication of being in such a limbo off and on for several days ahead of dying, and for quite a few days *after* death. An episode in which I was intimately and emotionally involved, her after-death appearance was remarkably bizarre, and I wanted very much to help, but had no idea how, other than meditating on the issue.

"The Park"

Robert Monroe's Experiments in the Afterlife

Since 1974, the Monroe Institute in Faber, Virginia, has offered programs exploring various aspects of mind and consciousness, particularly the lucid dreaming and afterlife states. The "rescue project" at the Institute claims to have found souls, minds, or psyches "stuck" in a non–temporal-spatial *replay* of the physical event (accident or violence), which ended their life. Because of this constant replay they fail to become aware of having died. The Institute's rescue squad interacts with such detached psyches, first explaining their dilemma to them, and then getting them involved in moving on by furnishing that isolated self a reciprocal point of awareness, sufficient to enable them to relate and indeed move on. I didn't place much stock in such dreamlike reports, which first struck me as illusions of genuine ego-enhancement on the part of the rescuers, but later saw how the general gist of the gesture-venture could occur after its own fashion.

Thousands of people have taken these Monroe training programs in subsequent decades, and, hardly incidental, the consensus that formed further strengthened such nonphysical field-effects and our possible entry into them. Aspects of that field apparently became stable and permanent, inherent within a common consensus. But we should remember that such "entry into fields" is a strange loop, and plays a part in bringing such field-effects about.

One memorable experience of Monroe's was finding himself in what he described as the Park, a tranquil field-state populated by souls or psyches of once-living people. In his training procedures for others, the Park became a milestone and indicator of progress, which thus became a commonly experienced event in those undergoing Monroe's training system, taking on greater permanence and strength (recall the mutual lucid dreaming of the young couple in San Francisco). And the more people who experienced the Park, the more stable and available that Park became—as with any field-effect.

Most significant of all, people discovered that in that Park they could "call up" the spirit of some departed person—generally someone quite close to them in life. Or, as it might be, some past but close person in their memory might be automatically called up by a previous relationship, when such opportunity arose (that is, those called up may have been tuned into and quite ready for such a call).

Since I live an easy walking-mile west of the Monroe Institute, I was visited by any number of enthusiastic, at times ecstatic people who had experienced not only an out-of-body state, but this very Park and the possibilities it afforded. A doctor friend from a nearby university hospital took the Monroe training program and found himself in the Park—and in the presence of his first great youthful love, killed in an auto accident at twenty-two years of age. The good doctor sat in my living room telling me about it soon afterward, weeping copiously over the impact of the encounter. He reported her as vividly alive, their conversation spontaneous and genuine. She was as real and lovable as he had known her, and looking about twenty-two, the age when she died (this aspect of looking the same age is a variable, and not necessarily a common factor). The good man came back time and again to take in the Monroe experience (originally a pricey eight-day marathon), drawn by his encounters.

Equally intriguing was a sixty-five-year-old globe-traveling engineer from Florida whose wife had died some six months previously. Unable to cope, the man was on the verge of emotional collapse when friends told him of Monroe's training, which he sought out and took. And indeed the engineer found himself in that Park, where he did indeed meet his wife, both rejoicing in their reunion, the good man weeping as heartily in recounting this to me as the doctor had, and also signing up for repeats of the course.

Dark Shadows

A further intriguing and disturbing aspect of Monroe's personal experiences, which he described at some length, was finding that immediately on leaving his body and world he encountered a demonic realm, a space or state filled with demonically driven homeless soul-mind-spirits who were terrifying, and literally tore at Monroe to pull him into their maelstrom of anger-hate and hellish makeup. Monroe managed to reject any trace of fear, hold his integrity of mind intact, and pass quickly through this dimension without harm. And I know of no reports among the thousands taking the Monroe program in the ensuing thirty-five or so years of participants having to experience this frightful passage, which, actual or not, in itself is indicative of field-effect.

This feature of a demonic realm, coupled with the problem of holding the mind's integrity on separation from the physical, might account for the practice by Tibetan lamas of gathering in a circle around the bedside of a dying brother and chanting continually for varying lengths of time. They may in effect pool their spiritual strength, and through their joint intent direct it toward the dying one to help him hold his own integrity of mind through that same or some similar barrier, as he goes through his transition from this world to the next.

This notion of demonic "disincarnates" also brings to mind reports of an incident in the closing hours of life of my great hero-teacher-guide, Baba Muktananda. Shortly before he left his body (for which departure he had carefully planned long in advance) he called in his trusted attendant Noni. He reported to Noni that three demons had just intruded on his presence, against which he had to summon his will and strength to dismiss. Seriously offended by such intrusion, Baba exclaimed to Noni, "What could those demons be doing here? How did they get into such a holy place?" (And Baba's ashram was indeed a holy place, it seemed to me, about the last place one would expect demons.) Following this encounter and report to Noni, Baba resumed his meditation posture and died quietly a few minutes later, having, in effect, cleared his path of possible debris. Such debris may lie deep within a person's memory, long since covered over and forgotten, but would, through "field-effect" attraction, be activated in our closing hours to clutter up the scene.

A final variation on this phenomena was given in the closing hours of my first wife's death, in her thirty-fifth year (near half a century ago), a process that extended over many hours and was as seriously moving and impressive to witness as it was shattering and heartbreaking to me. Throughout those last hours she was quite lucid, excited over Jesus sitting at the edge of her bed, while the room, she reported, filled with a brilliant gold light. At length she reported that Jesus, holding her hand, beckoned her to follow him, and she, grabbing my hand in turn, urged me to go with her into that golden light. (I, alas, in a heart too hardened and encased to see either Jesus or that gold light was left with only my massive pain at her approaching departure.)

At that point, however, she suddenly exclaimed, "I'm scared . . ." which was hardly in keeping with the rest of her experience. Whereupon she promptly dropped off into a brief, momentary sleep—but quickly awakened and reported that she had just dreamed that she was surrounded by a bunch of Baptists, who were fixing to pop her into a huge witch's cauldron. One would have to know her childhood history as an Episcopalian to understand why Baptists, buried somewhere deep in her unconscious, populating some dreamed demonic state, now briefly intruded on so beautiful a golden-lit realm. With that report to me, she quietly and peacefully drifted off again into a sleep from which she didn't awaken.

She made several "paranormal" appearances in the days following her leaving, all concerning our seriously damaged infant, born during my wife's last long illness, and the focus of her mind for weeks preceding her death. Two of these after-death events took place as I held the child, followed by two visible, and one sensory-felt visit she again made to her child—these latter at my wife's own mother's home, where her mother had taken the helpless grandchild to care for her.

On the distraught grandmother's vivid account of this final appearance-felt state, reported to me immediately after by telephone, I was reminded of my wife's encounter with her childhood memory of Baptists and such, which apparently had cleared her way shortly before leaving—as with Muktananda, a grace given for a grace-filled spirit.

As a final postscript here, I was reminded of a comment made by my greatest hero of some two millennia back, when, facing his own departure, he said, "I go to prepare a place for you," a simple ontological fact of our endless falling-forward, a "place" opening for my wife in those golden hours before she finally took his hand, there quite ready for her, at that last breath.

APPENDIX A

——◆——

MIND

The Last to Know

A Summary of HeartMath's
Original Research on the Precognitive Heart

A subject is wired up for both brain and heart electromagnetic activity (electrocardiograms and electroencephalograms), which is recorded and viewed by the person running the experiment, but not the subject. The subject sits in front of a separate monitor on which the computer will show images as selected for each run-through of the tests. When relaxed and ready for a trial, the subject presses a button that activates a ten-second timer, which, in turn will activate a random-selective device in the computer, which will then select and display on the screen a picture available to the subject being tested.

The pictures available for this random selection contain a mixture of pleasant and (seriously) unpleasant images. Ten seconds elapse between the subject pressing the button activating the timer and the selection actually made by the random-selector and displayed on the screen.

Anywhere from four to seven seconds before the selection is actually made by the timed random device, the electrocardiogram of the subject's heart shows a decisive shift in the heart's frequency pattern—

according to the nature of that upcoming image—which is *yet to be selected*. This shift of heart pattern activates a near-instant corresponding shift in the frontal brain, which then follows the heart pattern in the frequency action taking place. Note that the heart action precedes, if only by a fraction of a second, the following frontal brain action.

Of major significance here is that the heart-brain shift and resulting pattern of activity is decisively different when a positive image will be flashed on the screen at the end of the ten seconds, rather than when a negative image will be chosen. A negative image activates patterns indicative of a sympathetic nervous system's emotional-hormonal alert-reaction in the viewer (mild forms of the old "flight-fight" survival patterns in the "reptilian" sensory motor system). Positive precognitive images reinforce a parasympathetic nervous system's relaxed state of balanced repose. The viewing subject, however, is unaware of his or her heart-brain precognitive responses (occurring anywhere from four to seven seconds ahead), and is personally aware of and responsive to only the actual image itself when that image does flash forth on the screen, the ten-second trial thus completed.

An intuitive or precognitive capacity of the heart to "foretell" a future "machine-made" action is clearly indicated, which triggers the body-brain's emotional preparation for such action, if action is called for by negative information. Coming from those eventually visible results of an electromagnetically controlled laboratory effect, such intuitive precognition, as the heart clearly shows, is inexplicable within any physical process known. And, cautious about such an unknown enigma, before publishing their first paper on the anomaly, HeartMath ran 2,400 trials of the experiment—all of which verified the precognitive action in question.

As shown by the new heat-detecting brain monitors, when in the negative response pattern the bulk of brain activity shifts into a "hindbrain" defensive position, whereas positive images bring a balance of hind- and fore-brain synchrony in line with prefrontal cortex–heart influence.

This precognitive "alerting response" in heart and brain was an enigma unexpected by the research team. The research had originally grown out of experiments set up by Dean Radin and Karl Pribram to investigate galvanic skin response, also a long-standing enigma to physiologists.

As an intriguing aside, brought to my attention by my friend Robert Simmons, author of two recent books on crystals and their nature, our understanding of this enigma is beginning to unfold in the light of the research of Mae-Wan Ho at the Open University in London. Mae-Wan Ho discovered the crystalline nature of living cells and the resulting coherent crystalline state of our body. This discovery indicates that our body is a singular, whole crystalline state "immersed" in a singular (universal) matrix or frequency domain, from which matrix the various neural organizations making up our body select, according to the crystalline-based resonance making up those various organs of our body.

Paul MacLean first explored the role resonance plays in brain activity, which can now be linked to this crystalline state. This resonant selectivity body—made from a universal matrix of a crystalline nature—enters into the assembly or "putting together" of our sensory impressions of our world-to-view. This takes care of that old problem of galvanic skin response and will prove a major factor in a new and richer view of ourselves and creation that is in the making. Further, it can account for the various levels and forms of resonance sensed by and used in the "world-making" of preliterate societies such as the Kalahari !Kung, Australian Aborigine, Sng'oi, and others, whose world is so dramatically different from ours. Thank you, Robert Simmons—who sensed all this and much more in his two beautifully illustrated and brilliantly descriptive books.

Every bit as significant an aspect of this lengthy and involved experiment at HeartMath was strangely overlooked: the simple fact that none of the various subjects viewing that monitor for those 2,400 trials were aware of the precognitive maneuvers going on within their own heart-brain. Each subject's only conscious awareness was of the end-product:

the eventual image displayed on the screen itself; the subjects had no inkling of what was going on "behind the scenes" within their own conscious processes (an awareness that might, in fact, clutter up the cognitive awareness that is the point of it all—Nature knew what she was doing). Our knowledge of process is not necessary to respond intelligently to the *product* given us by that process—and our attempts to understand first and respond secondly lead to serious error.

This disconnect between the subject's conscious mind and the obvious conscious awareness of this heart-brain-sensory system, with its vast complexities, is a dramatic example of mind as an emergent property of the very body-brain-heart processes making mind possible, or giving the mind its sensory world experience, or both.

Letting-Allowing as Key

Since those interactions entering into, perhaps even bringing about, mind's awareness are not themselves available to that mind itself, we could say in this case that mind is aware only of the product, not the process. And in this regard we might propose that a learned task or discipline of the mind might be to *trust* process in order to engage in an ongoing interaction with products and the possibilities they afford.

This falls into line with poet Blake's observation that "mechanical excellence is the Vehicle for Genius." This stipulation may be a key factor in Rudolf Steiner's proposal that our great challenge is to allow the heart to teach us a new way of thinking, since through that new way of thinking the heart will find its own next stage of evolution.

Another Mind-Example:
An Education Conference in Hawaii, 2001

This aspect of our cognitive state, wherein wired-up subjects are themselves unaware of any disjunct between their personal mind-awareness and the complex of heart-brain precognitive processes giving rise to their awareness, was played out even more dramatically before an audience of some two hundred people at an educational conference in 2001.

A bright sixteen-year-old was wired up for brain-heart activity, the apparatus hooked into a computer which projected the results on a huge screen visible to the audience—but not to the young man, who was facing the audience, his back to the screen. The HeartMath representative running the experiment talked with the young man, getting him into so relaxed a state of mind that the young man's heart and brain went into "entrainment," which means they fell into the same synchronous or coherent electromagnetic wave patterns, an infrequent event for most of us, and clearly evident there on the audience's screen. The HeartMath representative, a British physician, congratulated the young man on his accomplishment, and turned him to look at the results on the screen, then back to the audience. The young man was beaming since he was familiar with HeartMath and knew the synchronous state was an accomplishment. At that point the doctor casually remarked that the young man was in such a coherent state he, the experimenter, wanted to try out a few orally presented mathematical problems used with young people in Britain.

Instantly when the mathematical testing was proposed, the young man's coherent state, visible to all but the young man, collapsed into an incoherent, chaotic state—similar, I might add, to the one making up or reflected in most of our "thinking" and emotional activity. The collapse of the young man's frequency-mapping display was so immediate and complete that we, the audience, burst into laughter. The young man looked puzzled and asked the HeartMath person, "What's up? What are people laughing about?" The good doctor then turned the young man to again view his now incoherent, scrambled heart-brain patterns, at which point the young man's face fell, as the saying goes.

The issue was clear: the emotional state of the "wired-up" subject had changed instantly and dramatically when the apparently threatening issue of being mathematically tested arose. The lesson for all of us educators and our educational endeavors was compelling: a student's availability to his or her own full-brain intellectual capacities is determined by his or her emotional state. Thus, the establishment of a nurturing, unthreatening ambient environment for students is the first and most primary principle for any learning situation. That is, whether or not the wired-up subject's prefrontal cortex–heart connections were fully operative or their attention-energy had been shifted into the ancient flight-fight defensive "hind-brain," limiting their intellectual capacity, could be easily observed. Thus a student, on being called on to perform—that is, be tested—may well have far less access to his or her own true level of intelligence, while unaware of being "cut off," in effect, from his or her own actual capacity, and left only with the feeling of inadequacy so many of us, students and teachers alike, feel.

Again, however, the major issue in this episode was neither recognized nor brought up again—the stark fact that the mind, being an "emergent aspect" of the brain, is not aware of that brain's actions, only the finally emerging cognitive affect-awareness, which can be rather "after the fact." Nor, I must insist, is the mind *supposed* to be so aware (at that time). There is a time for "left-hand thinking" and a time for "right-hand thinking."

This is an issue we must leave hanging here. But, along with the laboratory trials concerning emotional response and heart-brain shifts just mentioned, the disconnect between mind and brain-heart action clearly indicates that the mind is the recipient, not the principal causal factor of thinking and learning, as we generally assume. The mind, as I have said, is like the betrayed spouse—the last to know concerning creative thinking or clear reception of the overall cognitive process within us.

One might ask, then, who is in charge here? In this question we can get a glimpse of David Bohm's "thought as a system," and even the Eastern concept of *maya*—or play of illusions.

To train the mind to receive and translate selectively, rather than the idle meandering of our vague mental wanderings, was a serious issue with Rudolf Steiner. He urged us to differentiate between true creative-imaginative thinking—which demands discipline—and the idle wish-think of imaginary gratifications or disgruntled complaints churning forth in ordinary "roof-brain chatter" or daydreaming.

The time for idle imaginary-dreaming of the early child and the disciplined imagination of the adult must each have its day in the sun, however, since the latter is built on the former. And we should recall at this point why the mind must be suspended for the Laski Eureka response to function—which is again a matter of "not letting the right hand know what the left is doing." Since right-hand thinking is the mind as we ordinarily know it, and is critically necessary if we are to develop that Mechanical Excellence needed as the foundation for *allowing* that "left-hand thinking" of genius to flash forth, all of that behind-the-scene action, which we know not of, is action we do not *need* to know for the creative function to fulfill itself through us—for through our attempts to "know first" and experience secondly, we fall, time and again.

A BRIEF SUMMARY OF THE RESEARCH OF JAMES PRESCOTT AND MICHEL ODENT

James Prescott was for some sixteen years with the National Institutes of Health and their subsidiary, the Department of Health, Education, and Welfare. His specialty was early child development. In the late 1970s he and his associates were given a grant by the U.S. government to make the first scientific study of the root causes of crime and violence ever undertaken. Prescott and his team undertook the assignment with passion and vigor. Among many of their endeavors they set up their own "Primate Laboratory" to study the conception, birth, and growth of our ancient cousins, as well as following the research in other primate laboratories and field reports such as those of Jane Goodall and others. They exhaustively reviewed all current studies and research reports on birthing, the nurturing, and care of infants in a wide variety of societies, as well as studies of childhood-adolescent development in the United States and a host of both "primitive" and modern cultures, rich and poor.

In the final stages of their research they investigated anthropological studies and firsthand reports on fifty different cultures worldwide to determine the behavioral characteristics of so wide a spectrum: were the members of a society essentially benevolent or violent—both among themselves and in their interactions with other societies?

Their final summaries were presented in two sections, one concerning birth-bonding and early development, the other the treatment of adolescent populations from earliest puberty to adulthood. They were the first research group to show the clear parallels between the "toddler period" of early development and the period of late childhood and adolescence.

As covered in this present book, details of the two main compulsive drives of the toddler are to explore its new body-world, and build neural "structures of knowledge" accordingly. Prescott's studies showed that a nearly exact set of drives is found in adolescents. Given, in less than a year of astonishing growth, a new body, with new hormones, drives, forms of relationship demanding attention, and a seriously different reorganization of the brain, which will not be completed until the twenties, the adolescent is truly reborn into a new world on just as extensive and all-encompassing a level as the toddler, and must, if we are to survive, be treated with the same nurturing care and consideration we are destined by Nature to give to our infant-children (whether we do or not).

If a society is nurturing and benevolent toward its infants in birth, bonding, and early growth, that society will create and maintain about half the requirements for a benevolent, caring cultural ambient. If a society is nurturing, benevolent, tolerant, and caring for its adolescent population, that society will likewise develop about half the requirements and distinguishing behaviors of a benevolent, caring society. A society failing to nurture their young, yet lenient toward their adolescents and their extremities of behaviors (particularly sexual), will be partially benevolent toward their own kind. If a society nurtures its young but is repressive toward its adolescents, it will be about half-tolerant and half-benevolent among its members.

Any society that is both unnurturing and repressive toward its infant-young *and* toward its adolescents will be hostile and violent within its own border and kin, and hostile or warlike toward its neighbors. There is a direct, one-for-one correspondence.

The United States, Prescott's HEW report stated, fails on both counts—abandoning and failing to nurture its young, while harsh and repressive toward its adolescents, particularly in regard to sexuality.

In regard to early infant nurturing, Prescott's group found that mothers in nurturing societies carry their infants for at least the first year of life. This action provides the infant mind-brain-body a most critical factor—the nurturing effect of sufficient movement of the body and close contact with a caretaker. Frequent body movement, whether through the nurturing caretaker or self-stimuli by the infant (in their own near constant movement of body when awake) and being moved as when carried, is critical to the growth and development of the cerebellum, the entire sensory-motor system (including vision, balance and orientation, hearing and speech), and a laundry list of related aspects of development that will affect the individual lifelong.

"Touch deprivation" is rampant in American children, brought about by failure of sufficient movement, holding, touching and general sensory-stimulation in the early years. One thing never experienced in the womb is stillness or silence. From conception, the embryo, fetus, and infant is in some continual form of motion and sound—the mother's breathing, heartbeat, blood circulation, speech, as well as any movement of the mother herself. Born in the sterile world of the hospital, isolated in cubicles, seldom picked up, deprivation occurs from the beginning. Spending their earliest and most critical sensory-development periods in confining cribs, bassinets, carriages, strollers—all isolating the child in a touch-deprived environment—brings impairment on most of the levels of sensory experience, which brain development and communication must have. Touch-starvation in American children, as reported by French physician Alfred Tomatis, has a direct corollary in speech and hearing deficiencies. Touch-deprived children demand and seek greater volumes of sound, leading to intense, ever-increasing decibel levels of the music they are attracted to, greater levels of direct, close-packed physical contacts and less and less capacity for silence or concentration and learning, particularly on abstract levels. Meanwhile autism, in its varying expressions, increases.

James Prescott, having provided extensive evidence for how and why all this takes place, knew serious frustration over having been largely excluded from his own profession by having suggested, in his massive study of adolescents, sexual freedom for those adolescents, as found in all benevolent societies. The religious right had just taken over under Ronald Reagan when Prescott submitted his works for approval by the administrators of the NIH. Like Candace Pert, who was blackballed by NIH for challenging their authority, Prescott knew growing frustration over the years, watching the rising tide of dysfunction found at every level of American children, all in the face of the massive evidence he had collected concerning the cause. Michael Mendizza (Touch the Future, www.ttfuture.org) has cataloged most of Prescott's work, making it available today, when it is more pertinent to our needs than ever.

Michel Odent

French physician Michel Odent revolutionized birthing at Pithiviers Hospital in France, and eventually set up a research clinic in London for a full-dimension physiological analysis of all elements of sexuality, conception, birthing, breastfeeding, and general care of infants. During his tutelage at Pithiviers, his department delivered nearly 3,000 infants without loss of a mother or child.

The two works produced to date from Odent's London research center are *Scientification of Love* and *The Functions of the Orgasms*, which, combined, hold within them not only "the highway to transcendence" (as Odent called it), but that which could be the very saving grace for our species as a whole. In no way can I overstate the importance of Odent's works, and in nothing less than a whole book could one convey the vast wealth of critical information and observations offered in Odent's simple, easily grasped explanations. But I will try here to give some scant sampling of Odent's invaluable offering and knowledge—some known for ages, lost time and again and rediscovered time and again—all now backed by true scientific studies of physiology. Bear in mind Maria Montessori's prophetic warnings of nearly a century ago and quoted often in this present book, that humankind abandoned in its earliest formative period becomes its own greatest threat to survival, and add to this Charles Darwin's studies showing our species was brought about through love and altruism. It is time we defined those terms and truly embraced them in our conscious awareness. Following are some samples from Odent:

1. The capacity to love is determined, to a great extent, by the early experience in fetal life and in the period surrounding birth.
2. A study of how we learn to love starts at the breast a few seconds after birth and holds a clue to the cause of violence or peace in our society.
3. Odent proposes that the capacity to love is encapsulated in the not very complex molecule of oxytocin, which holds the capacity to love and protect our planet, the prerequisite of global survival.
4. Today the nature of love and how the capacity to love develops has become a subject for scientific study, the implications of which are at least as important as those of genetics, electronics, or quantum theory.
5. For all the different manifestations of love—maternal, paternal, filial, sexual, romantic, platonic, spiritual, brotherly, not to mention love of country, love of inanimate objects, compassion and concern for Mother Earth—the prototype is maternal love.
6. Evidence points to a very short and yet critical period of time just after birth, which has long-term consequences so far as our future capacity to love is concerned. We disregard the consequences of interfering with or neglecting the physiology of that critical period at our peril.

7. The biological sciences represent a mirror in which we look for a reflection of ourselves. Today, that mirror has been brilliantly polished and broken into a thousand pieces. Our objective must be to establish links between all those pieces.

8. The different hormones released by the mother and baby during labor and delivery all play a specific role in the ongoing interactions between mother and baby.

9. The greater the social compulsions for aggression and destruction of life, the more intrusive relations become between infant and mother, through the imposed cultural rituals and beliefs in regard to the period surrounding birth.

10. The relationship with the birth-mother and the relationship with Mother Earth are two aspects of the same phenomenon.

11. The language of modern physiologists can clearly explain what is happening when a woman is giving birth.

12. When a woman is giving birth, the most active part of her body is the ancient primitive brain (reptilian-sensory-motor).

13. When there are inhibitions in giving birth or any other sexual activity, the developed neo-cortex (latest evolutionary brain) is activated and in charge (which is an exact turnaround of the developmental roles of hind-brain and fore-brain, detailed in this present work).

14. Any neocortical stimulation, particularly of the intellect, will interfere with the process of delivery.

15. All mammals have a strategy for giving birth in privacy, generally in the dark.

16. Among more natural societies, as in most mammalian life, where women prefer to birth in solitude, even the snapping of a twig indicates possible intrusion and can slow or halt the birth process, waiting for "the coast to clear." Precisely those prerequisites for quiet and safety still hold in the modern woman—just massively ignored and overridden by astonishing medical intrusions throughout the birth process, in the best of situations. (Thus,

lengthy, painful, and dangerous birthing becomes the norm. Intellect replaces intelligence and compassion. Home birth is far safer for both mother and infant than any hospital situation.)

17. A laboring woman needs to feel, above all, safe and secure; this is a prerequisite for the changing level of her consciousness, which is a characteristic of the natural birth process.

18. There is a current cultural misunderstanding of the birth process. This misunderstanding is transmitted by nonverbal messages, as when books concerning birth show the laboring woman surrounded by anxious people watching her, advising, reporting progress, all suggesting impending danger and the mother's needs that require their help in her labors—all delaying and making labor more and more difficult, and dangerous.

19. Researchers have found the only significant result of electronic monitoring of birth, now common practice in most obstetrical situations, is an increase in cesarean sectioning (which is far more lucrative to both doctor and hospital).

20. Any stimulation of the mother's neo-cortex—talking to her, surrounding her with bright lights, making her feel observed, insecure or unsure, or any action stimulating her release of adrenaline—inhibits the birth process.

21. Sexual intercourse, childbirth, and lactation can be inhibited by the same neocortical centers—or neocortical "brakes." Modern physiologists view sexuality as a whole. Intercourse bringing pregnancy, and birthing completing it, are twin functions in Nature's way.

22. Oxytocin is one of the main hormones involved in the different aspect of male and female sexuality.

23. Oxytocin is the hormone capable of inducing maternal behavior in the hour following birth. (Marshall Klaus, of Case Western Reserve University, said there was about a forty-five-minute window open for establishing the critical bonds between infant and mother—after that, such bonding is difficult, insufficient,

even unavailable, abandonment then being complete and near irrevocable.)

24. By releasing its own oxytocin, a fetus may contribute to its own delivery.

25. During intercourse, childbirth, and lactation, two groups of hormones play a pre-eminent role—the altruistic hormone oxytocin and the beta endorphins that can be considered our "reward" systems. An integrated vision of sexual life inspired by modern biological science has practical implications for our survival.

26. Beta endorphins, akin to oxytocin in ability to bring peaks or orgasms, occur during labor and peak as well during breastfeeding. Such hormones play a critical role in the ongoing development of the infant and well-being of the mother. (Recall the report from Israel several years ago that any society eliminating or curtailing breastfeeding has an immediate, one-for-one corresponding increase in breast cancer—information never published in the United States and, ironically, ignored by most Israeli doctors as well, who are as blinded by their own ego-inflation and economic interests as much as American physicians.)

27. The senses of taste, smell, and hearing play an important role in the baby's identification of its mother before and after birth.

This list can go on and on, and grow quite technical. So at this point I would again quote Maria Montessori in her observation that humankind abandoned in its earliest formative period becomes its own worst threat to its survival. Doctors, midwives, hospitals, and helping friends all inadvertently feed on, and in turn feed into, a laboring mother's feelings of inadequacy and helplessness, the various enemies within. In this regard, a bit of knowledge of recent biological research would enlighten, and is needed on all levels of our society, most seriously by the delivering mother, who should be protected by a new standard-bearer concerning the world around that mother: "Keep laboring women out of reach of meddling members of society around them." Solo birthing is being discovered by more and more courageous, intelligent women, and is paying remarkable dividends.

One of the great barriers to women regaining control over their own lives, and their birthing and mothering capacities is the fearful set of obstructions men and fathers bring about. The notion that fathers are important at birthing is a new myth, and is not the case. Women have handled these issues for millennia—a power still genetically encoded, ready to go to work.

NOTES

Introduction. A Mirror of the Universe

1. Margulis and Sagan, *Microcosmos.*
2. Ho, *The Rainbow and the Worm.*

Chapter 1. The Fall of Man

1. See Wolff, *Original Wisdom;* and Leidloff, *The Continuum Concept.*
2. Goodall, *The Chimpanzees of Gombe.*
3. Wilson, *The Hand: How Its Use Shapes the Brain, Language and Human Culture.*

Chapter 2. Emotion in Evolution

1. This article's provenance is now lost, but subsequent research by Bruce Lipton document the same phenomenon. See Lipton, *The Biology of Belief.*
2. Pert, *Molecules of Emotion.*
3. Schore, *Affect Regulation and the Origin of the Self.*

Chapter 3. The Great Conflict

1. Miller, *For Your Own Good.*
2. Leidloff, *The Continuum Concept.*
3. Childre and Martin, *The HeartMath Solution;* see also www.heartmath.com.

Chapter 4. Cultural Default and Intentional Evolution

1. Pearce, *The Biology of Transcendence.*

Chapter 6. Scientific Perspectives of
Mind-Heart and Resonant Fields

1. Pert, *Molecules of Emotion.*
2. Childre and Martin, *The HeartMath Solution* and www.heartmath.com.
3. Buzzell, *Children of Cyclops.*
4. Sperry, "Mind-Brain Interactions," 193–206.

Chapter 7. Nature's Plan and Culture's Conniving

1. For an extensive treatment of the primacy of movement, see "The Art of Perceiving Movement" in Bohm, *On Creativity.*
2. Williamson and Pearse, *Science, Synthesis and Society.*

Chapter 8. Darwin's Evolution

1. Storpher, *Intelligence and Giftedness.*
2. Montagu, *The Natural Superiority of Women,* and Storpher's classic study *Intelligence and Giftedness* offer clarification of these issues, which lie quite beyond any petty "gender warfare" or conflict, as plagues "civilized" humans today.

Chapter 9. Darwin II: Death and the Evolution of the Mind

1. Weber, *Dialogues with Scientists and Sages.*
2. Laski, *Ecstasy in Secular and Religious Experience.*

Chapter 10. Mind, Spirit, and Creative Fields

1. Gardner, *Frames of Mind.*
2. MacLean, *The Triune Brain in Evolution.*
3. Von Halle, *And If He Has Not Been Raised.*
4. Werner, *Life from Light.*
5. Werner, "Professor claims to survive on just sunshine and fruit juice," *UK Daily Mail,* June 28, 2007.

BIBLIOGRAPHY

Bohm, David. "The Art of Perceiving Movement." *On Creativity*. London: Routledge, 1998.

Buzzell, Keith. *Children of Cyclops*. Association of Waldorf Schools of North America Publications, 1998.

Childre, Doc Lew, and Howard Martin. *The HeartMath Solution*. New York: HarperOne, 2000.

Gardner, Howard. *Frames of Mind: The Theory of Multiple Intelligences*. New York: Basic Books, 1993.

Goodall, Jane. *The Chimpanzees of Gombe*. Cambridge, Mass.: Belknap Press, 1986.

Halle, Judith von. *And If He Has Not Been Raised: The Stations of Christ's Path to Spirit Man*. East Sussex: Temple Lodge Publishing, 2007.

Ho, Mae-Wan. *The Rainbow and the Worm: The Physics of Organisms*. New Jersey: World Scientific Publishing, 2008.

Jawer, Michael A., and Marc S. Micozzi. *The Spiritual Anatomy of Emotion: How Feelings Link the Brain, the Body, and the Sixth Sense*. Rochester, Vt.: Park Street Press, 2009.

Laski, Marghanita. *Ecstasy in Secular and Religious Experience*. New York: JP Tarcher, 1990.

Leidloff, Jean. *The Continuum Concept*. Reading, Mass.: Addison Wesley Press, 1977.

Lipton, Bruce. *The Biology of Belief*. Carlsbad Calif.: Hay House Publishing, 2008.

MacLean, Paul. *The Triune Brain in Evolution: Role in Paleocerebral Functions*. New York: Springer Publishing, 1990.

Margulis, Lynn, and Dorian Sagan. *Microcosmos: Four Billion Years of Microbial Evolution.* Berkeley: University of California Press, 1997.

Miller, Alice. *For Your Own Good: Hidden Cruelty in Child-Rearing and the Roots of Violence.* New York: Farrar, Straus, and Giroux, 1990.

Montagu, Ashley. *The Natural Superiority of Women.* New York: Collier Books, 1992.

Pearce, Joseph Chilton. *The Biology of Transcendence: A Blueprint of the Human Spirit.* Rochester, Vt.: Park Street Press, 2004.

———. *The Crack in the Cosmic Egg.* Rochester, Vt.: Park Street Press, 2002.

Pert, Candace. *Molecules of Emotion: The Science behind Mind-Body Medicine.* New York: Simon and Schuster, 1999.

Schore, Allan N. *Affect Regulation and the Origin of the Self: The Neurobiology of Emotional Development.* Abingdon, UK: Psychology Press, 1999.

Sperry, Roger. "Mind-Brain Interactions," *Neuroscience* 5, 193–206.

Storpher, Miles. *Intelligence and Giftedness: The Contributions of Heredity and Early Environment.* San Francisco: Jossey-Bass, 1990.

Weber, Renee. *Dialogues with Scientists and Sages.* London: Arkana, 1990.

Werner, Michael. *Life from Light.* East Sussex: Clairview Books, 2007.

Williamson, G. Scott, and Innes Hope Pearse. *Science, Synthesis and Society.* Jedburgh, Scotland: Scottish Academic Press, 1980.

Wilson, Frank. *The Hand: How Its Use Shapes the Brain, Language and Human Culture.* London: Vintage Books, 1999.

Wolff, Robert. *Original Wisdom.* Rochester, Vt.: Inner Traditions, 2002.

INDEX

Page numbers in *italics* represent illustrations.

Darwin's Lost Theory of Love, 10, 111–12
Davenport, Marcia, 122–24
death, xiv–xv, 3–4, 98, 128
demonic realm, 176–78
Department of Health, Education, and
    Welfare (HEW), 53
Descent of Man, The. See Darwin II
distortion, 147
DNA, 5–6, 28–29, 104–5
    form and content of, 103–4
    understanding of, 101–3
Dreamtime, 126

Earth, 79–80, 95–96, 97
Eckhart, Meister, 99, 140, 159
Ecstasy, 116
Eddington, Arthur, 1
Edleman, Gerald, 137
eggs, 105–7
electricity, 121–22
electrocardiograms, 68, 70
electromagnetic fields, 70–74, *71, 73,*
    77–79
Eliade, Mircea, 155
El Shaddai, 156
embryogenesis, 104–7
emotional fields, evolution and, 51–53
emotional intelligence, 24–25
emotions, 3–4, 31–34, 76–77
Eucharist, 146
Eureka! experience, 116, 118–20
evolution, xv, 1
    as aspect of creation, 2–3
    Darwin and, 112
    death and, 3–4, 98
    emotional fields and, 51–53

mitochondria and, 4–6
strange loop of, 98–100
unfolding of, 173–75
work-in-progress of, 130–31
Evolution's End, 57–58
exploration, 39–40, 43, 44–45
"Expressions of the Emotions in Man
    and Animals," 111–12
eye contact, 43

fall of man, 24–25, 26
fasting, 141–43
fear, 3–4
field-effects, xii, 127–28, 136–37,
    140–43
field phenomenon, 115–17, 120–21,
    124–25
field resonance
    in childhood, 124–25
    in indigenous culture, 125–27
    in time, 133–36
    von Halle and, 136–40
First Three Years, The, 124–25
flight-fight hormones, 82
Formal Operations, 20
Fox, George, 151, 159–60
Frances, St., 91, 138
Freud, Sigmund, 98
Functions of the Orgasm, The, 108,
    189

Gardner, Howard, 127, 135–36
genetic blueprints, 28–30
Gilligan, Carol, 147
God, 99, 152–53, 155–58
"God of Abraham, The," 154

BOOKS OF RELATED INTEREST

The Biology of Transcendence
A Blueprint of the Human Spirit
by Joseph Chilton Pearce

The Crack in the Cosmic Egg
New Constructs of Mind and Reality
by Joseph Chilton Pearce

The Death of Religion and the Rebirth of Spirit
A Return to the Intelligence of the Heart
by Joseph Chilton Pearce

Where Does Mind End?
A Radical History of Consciousness and the Awakened Self
by Marc Seifer, Ph.D.

New World Mindfulness
From the Founding Fathers, Emerson, and Thoreau
to Your Personal Practice
by Donald McCown and Marc S. Micozzi, M.D., Ph.D.

The Spiritual Anatomy of Emotion
How Feelings Link the Brain, the Body, and the Sixth Sense
by Michael A. Jawer with Marc S. Micozzi, M.D., Ph.D.

New Consciousness for a New World
How to Thrive in Transitional Times and
Participate in the Coming Spiritual Renaissance
by Kingsley L. Dennis

The Struggle for Your Mind
Conscious Evolution and the Battle to Control How We Think
by Kingsley L. Dennis

INNER TRADITIONS • BEAR & COMPANY
P.O. Box 388
Rochester, VT 05767
1-800-246-8648
www.InnerTraditions.com

Or contact your local bookseller